Impurity Transport in Magnetically Confined Plasmas

Online at: https://doi.org/10.1088/978-0-7503-1451-0

IOP Series in Plasma Physics

Series Editor

Richard Dendy

Culham Centre for Fusion Energy and the University of Warwick, UK

About the series

The IOP Plasma Physics ebook series aims at comprehensive coverage of the physics and applications of natural and laboratory plasmas, across all temperature regimes. Books in the series range from graduate and upper-level undergraduate textbooks, research monographs and reviews.

The conceptual areas of plasma physics addressed in the series include:

- Equilibrium, stability, and control.
- Waves: fundamental properties, emission, and absorption.
- Nonlinear phenomena and turbulence.
- Transport theory and phenomenology.
- Laser–plasma interactions.
- Non-thermal and suprathermal particle populations.
- Beams and non-neutral plasmas.
- High energy density physics.
- Plasma–solid interactions, dusty, complex, and non-ideal plasmas.
- Diagnostic measurements and techniques for data analysis.

The fields of application include:

- Nuclear fusion through magnetic and inertial confinement.
- Solar–terrestrial and astrophysical plasma environments and phenomena.
- Advanced radiation sources.
- Materials processing and functionalisation.
- Propulsion, combustion, and bulk materials management.
- Interaction of plasma with living matter and liquids.
- Biological, medical, and environmental systems.
- Low-temperature plasmas, glow discharges, and vacuum arcs.
- Plasma chemistry and reaction mechanisms.
- Plasma production by novel means.

A full list of titles published in this series can be found here: https://iopscience.iop.org/bookListInfo/iop-plasma-physics-series.

Impurity Transport in Magnetically Confined Plasmas

Katsumi Ida
Fusion Science Interdisciplinary Coordination Center/Department of Research, National Institute for Fusion Science, Toki, Japan

Naoki Tamura
Department of Research, National Institute for Fusion Science, Graduate Institute for Advanced Studies, SOKENDAI, Toki, Japan

IOP Publishing, Bristol, UK

ISBN 978-0-7503-1451-0 (ebook)
ISBN 978-0-7503-1449-7 (print)
ISBN 978-0-7503-1895-2 (myPrint)
ISBN 978-0-7503-1450-3 (mobi)

DOI 10.1088/978-0-7503-1451-0

Version: 20231201

IOP ebooks

British Library Cataloguing-in-Publication Data: A catalogue record for this book is available from the British Library.

Published by IOP Publishing, wholly owned by The Institute of Physics, London

IOP Publishing, No.2 The Distillery, Glassfields, Avon Street, Bristol, BS2 0GR, UK

US Office: IOP Publishing, Inc., 190 North Independence Mall West, Suite 601, Philadelphia, PA 19106, USA

Contents

Preface

This monograph is motivated by a renewed interest in impurity transport in nuclear fusion research. Before 1980, when plasma-facing materials were metal, intense radiation in the plasma core due to the accumulation of impurities was a serious problem in magnetically confined plasma. In order to solve this problem, low-Z plasma-facing materials, such as carbon, boron, and beryllium, were employed, and they have been used for three decades since. Since low-Z impurities do not cause a severe radiation problem in the plasma core, or even contribute to improving plasma confinement, impurity transport has become less mentioned in nuclear fusion research. However, the absorption of tritium in low-Z plasma-facing materials is a new problem for devices burning plasma. Therefore, the use of materials facing high-Z plasmas to reduce tritium retention has become mandatory, and as such the reduction and control of high-Z impurities is a new challenge in fusion research.

This monograph covers the diagnostics, experimental approach, and results of recent impurity-transport studies in tokamak and helical plasmas. It also covers the impurity transport parallel to the magnetic field in the scrape-off layer (SOL) and the impact of magnetic topology on impurity transport. The book is organized as follows. Chapter 1 provides an introduction to impurity transport, with an emphasis on the history of plasma-facing materials in tokamak and helical devices. Chapter 2 describes the impurity-transport model, focusing on the atomic and transport processes in closed and open flux surfaces. Chapter 3 discusses the diagnostics system with passive and active spectroscopy. Chapter 4 presents an experimental approach to impurity-transport study with intrinsic and externally injected impurities. Chapter 5 discusses impurity transport across magnetic flux surfaces, focusing on impurity accumulation, poloidal asymmetry, and the impurity hole. Chapter 6 presents impurity transport at the plasma edge and SOL, including the impurity source and transport parallel to the magnetic field. Chapter 7 describes the effect of magnetic topology on impurity transport, especially magnetic islands, stochastic field regions, and the last closed flux surface. Chapter 8 discusses the control of impurity transport using electron cyclotron resonance heating and ion cyclotron resonance heating.

Katsumi Ida and Naoki Tamura
National Institute for Fusion Science, Toki, Gifu
September 2023

Acknowledgements

We especially thank Shigeru Sudo, Professor Emeritus at the National Institute for Fusion Science, for encouraging us to write this monograph. We thank Dr Masahiro Kobayashi (National Institute for Fusion Science) for his assistance in preparing chapter 6. We also thank the Large Helical Device (LHD) experiment group at the National Institute for Fusion Science for their assistance in this research.

Author biographies

Katsumi Ida

Katsumi Ida received his BSc and MSc from The University of Tokyo in 1980 and 1982. Then, he went to Princeton University and worked under the supervision of Professor Raymond Fonck in 1984–85. He received a PhD from The University of Tokyo in 1986. He joined the faculty of Nagoya University in 1986 and the National Institute for Fusion Science in 1989. He has pioneered a new frontier in experimental studies of turbulent transport in toroidal plasmas and discovered many essential processes in the turbulent transport of magnetically confined plasmas in far non-equilibrium states. His discoveries have enhanced the value of plasma physics as a modern physics discipline and significantly contributed to the realization of fusion reactors. He discovered experimentally that an intrinsic toroidal flow could be created. He made explicit the essential aspect of turbulent plasma transport in which temperature gradients give rise to momentum flow. Actualizing these achievements, he has also significantly contributed to the realization of fusion reactors, as plasma rotation plays an essential role in steady-state tokamak reactors and in suppressing turbulence. He also received the Nishina Memorial Prize in 2011, Subrahmanyan Chandrasekhar Prize of Plasma Physics in 2023, and the entire physics community has widely recognized plasma physics. He has published over 500 papers, including 100 first-author papers in scientific journals, most of which are about turbulence and transport using spectroscopic measurements of impurity line emissions in toroidal plasmas.

Naoki Tamura

Naoki Tamura received his BE and ME from Nagoya University in 1997 and 1999, respectively. He then moved to the Graduate University for Advanced Studies (SOKENDAI), where he began research on impurity transport in magnetically confined toroidal plasmas, for which he obtained his PhD thesis. In this work, under the guidance of Professor Shigeru Sudo of SOKENDAI, he developed a tracer-encapsulated solid pellet (TESPEL) with the help of many international collaborators. The experience at that time made international collaborations essential to his subsequent research activities. TESPELs are used in various magnetically confined toroidal plasma experiment devices worldwide. It will also be used in JT-60SA, the world's largest superconducting tokamak device except ITER, which is scheduled to be fully operational soon. TESPELs have been of great benefit in their original purpose of impurity transport and transient heat transport. A significant result in the study of transient heat transport using TESPELs has been the discovery of the non-local heat transport phenomenon in stellarator plasmas. After commencing research on transient heat transport, he became very interested in the turbulent properties of plasmas, which are non-equilibrium open systems. Impurities hold the key to the success or failure of fusion reactors, and turbulences, a peculiar property of high-temperature plasma, have kept him captivated in his research, day and night, with the help of many overseas collaborators and friends.

Chapter 1

Introduction

The history of impurity control and materials in plasma-facing components (PFCs) are summarized in this chapter. The materials of PFCs were changed from high-Z materials (metals) to low-Z materials (carbon or boron) in order to reduce core radiation loss due to the accumulation of high-Z impurities. Although the radiation problem was solved by low-Z PFCs, these low-Z materials are expected to absorb the tritium in deuterium–tritium fusion (D-T) devices and cause a serious problem of tritium retention. In contrast, high-Z PFCs have much less tritium retention. As such, the merit of high-Z materials has again been recognized and PFCs have reverted back to utilizing high-Z materials.

1.1 High-Z impurity accumulation

The selection of the materials that will face the plasma is an important issue in impurity control in fusion devices. Previously, the reduction of radiation loss in the plasma core was an essential focus of fusion research in the 1980s [1]. High-Z impurities such as metals are only partially ionized even in the plasma core and cause a cooling of the plasma via significant radiation loss in the plasma. In contrast, low-Z impurities such as beryllium, boron, and carbon are fully ionized in the plasma core and do not cause the cooling of the plasma by radiation loss, but instead reduce the density of bulk ions in respect to the electron density, which is called dilution. As a heat flux to the facing materials (vessel wall, limiter, and divertor), high-Z impurities (metals) have been recognized to cause significant radiation loss from the plasma core, because such impurities tend to accumulate in the plasma center. In some cases, the strong radial loss causes a hollow electron temperature profile and deterioration of the plasma performance.

Figure 1.1 shows radial profiles of the electron temperature and density, measured with Thomson scattering, and the integrated radiation power, measured with a bolometer, and ohmic input power for tungsten limiters and graphite limiters in the Princeton Large Torus (PLT) [2, 3] and Impurity Study Experiment (ISX-B) [4]

doi:10.1088/978-0-7503-1451-0ch1

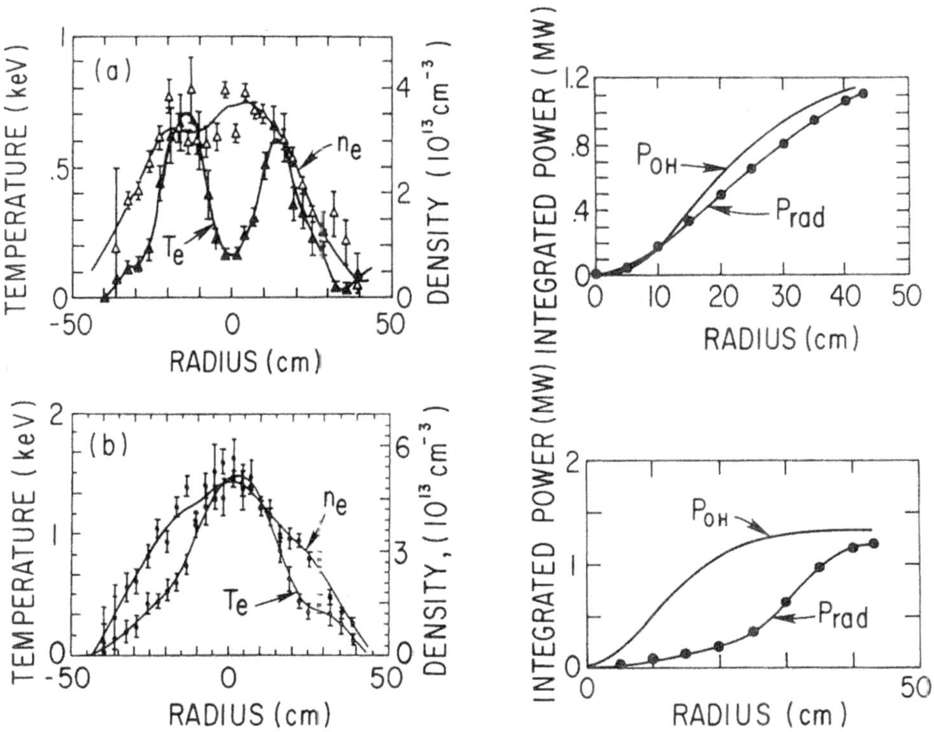

Figure 1.1. Profiles of electron temperature and density, measured with Thomson scattering, and integrated radiation power, measured with a bolometer, and ohmic input power for (a) tungsten limiters and (b) graphite limiters in PLT ohmic discharge. Reprinted from [2], Copyright (1980), with permission from Elsevier.

tokamaks. As can be seen in the case of tungsten limiters, the plasma performance is completely limited by the radiation. The electron temperature profile becomes hollow due to cooling by the significant radiation power, which slightly exceeds the ohmic input power. In contrast, when the central radiation is reduced to a level much lower than the ohmic input power, the electron temperature becomes peaked. Although the total radiation power is still high, ~90%, the peak of the radiation power shifts to near the plasma periphery. It was found that the central peaked radiation due to impurity accumulation can be avoided by changing the limiter materials from tungsten to carbon in ohmically heated plasma. However, the switching of the limiter from a high-Z material to a low-Z material is not sufficient to avoid the accumulation of impurities when the heating power is increased by additional heating processes such as neutral-beam injection (NBI) heating and ion cyclotron resonance heating (ICRH). Impurity accumulation and high central radiation present serious problems in high-confinement-mode (H-mode) plasmas.

Figure 1.2 shows the time evolution of line radiation of the iron line in co-injection (solid lines) and counter-injection (dashed lines) discharges in deuterium in

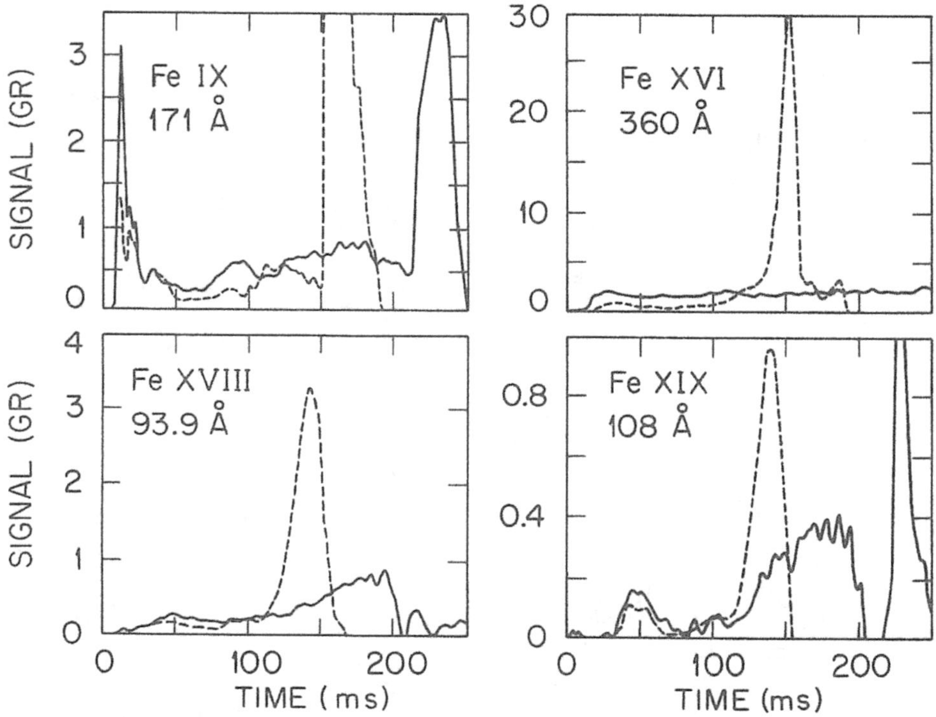

Figure 1.2. Time evolution of line radiation of the iron line in co-injection (solid lines) and counter-injection (dashed lines) discharges in deuterium in the ISX-B tokamak. Reprinted with permission from [5], Copyright (1981), by the American Physical Society.

the ISX-B tokamak [5]. A neutral beam is injected from 100 to 200 ms. As can be seen, the iron-line radiation with a lower charge state (Fe IX and Fe XVI) is almost unchanged through the discharge process, but the iron-line radiation with a higher charge state (Fe XVIII and Fe XIX) increases gradually after the NBI in the co-injection. In contrast, the iron-line radiation with a higher charge state (Fe XVIII and Fe XIX) sharply increases immediately after the NBI in the counter-injection due to the impurity accumulation. The simultaneous decrease of iron-line radiation with a higher charge state (Fe XVIII and Fe XIX) and increase of iron-line radiation with a lower charge state (Fe IX and Fe XVI) after 140 ms in the counter NBI are due to the recombination associated with a drop of central electron temperature. The sawtooth oscillation, which is predicted to inhibit impurity accumulation, disappears in the counter-injection NBI discharge. The disappearance of this oscillation accelerates the impurity accumulation and causes termination before the end of the counter NBI pulse.

Impurity accumulation occurs again after the transition from a low-confinement mode (L-mode) to a high-confinement mode (H-mode). This impurity accumulation can be mitigated by density control with a gas puff, which was found in ohmically heated plasma in the PLT [3]. A similar technique was applied to suppress impurity accumulation in the H-mode phase in the Princeton Beta Experiment (PBX)

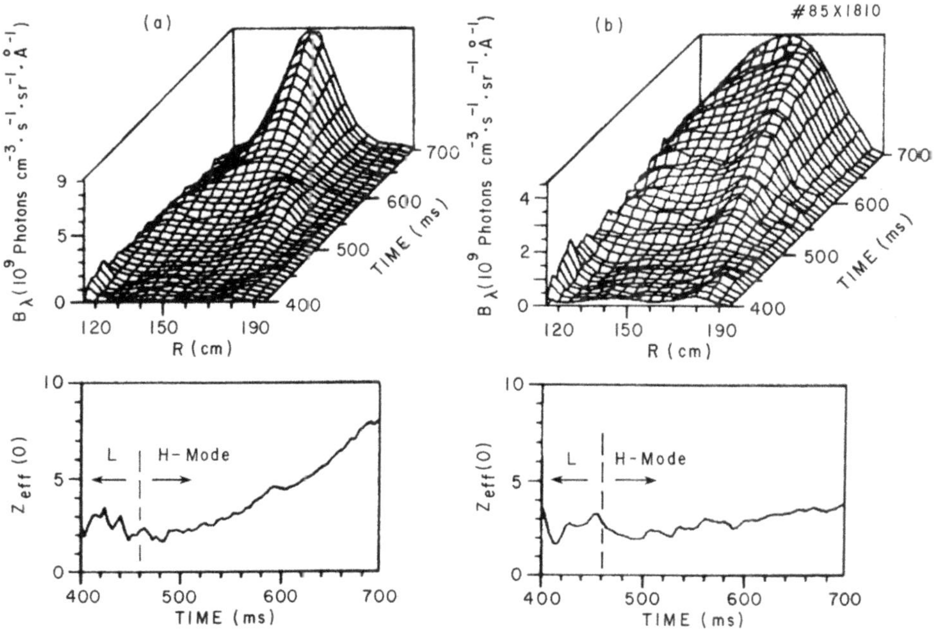

Figure 1.3. Time evolution of the visible bremsstrahlung profiles and the central Z_{eff} for (a) a discharge with no gas puff during NBI, and for (b) a discharge with a moderate gas puff after 400 ms in PBX H-mode discharge. Reproduced from [6]. © International Atomic Energy Agency. Published by IOP Publishing. All rights reserved.

tokamak, as can be seen in figure 1.3 [6]. The intensity of the visible bremsstrahlung is commonly used to estimate the $Z_{\text{eff}}[=\Sigma_i n_i Z_i^2/n_e]$, because its intensity is proportional to Z_{eff}, n_e^2, and $T_e^{-1/2}$. The radial profile of the intensity of the visible bremsstrahlung becomes very peaked when there is no gas puff after the transition from L-mode to H-mode (at $t = 460$ ms), and the central Z_{eff} sharply increases up to 8. The high central Z_{eff} clearly indicates the accumulation of impurities in the H-mode phase. This impurity accumulation can be suppressed by a gas puff after 400 ms, as demonstrated by the broad profile of the intensity of the visible bremsstrahlung. The central Z_{eff} gradually increases but is held below 4. Impurity accumulation can be suppressed by reducing the impurity influx by a moderate gas puff. These experiments suggest the non-linear character of the impurity accumulation process, where the inward convection velocity of the impurity increases as the concentration of the impurity is increased due to the collision process between impurities [6]. Although impurity accumulation can be avoided, impurity accumulation is always a serious problem in tokamak devices with high-Z plasma-facing components (PFCs).

1.2 Carbonization and boronization

In order to reduce radiation loss in the plasma core, several approaches to change the plasma-facing materials from high-Z materials to low-Z materials have been tested. Wall coating with a low-Z material (carbon or boron) was proposed and tested in various devices. Figure 1.4 shows schematic sketches of the carbonization technique and operation regime of the Tokamak Experiment for Technology Oriented Research (TEXTOR) with ohmic heating for conventional glow discharge cleaning and for carbonized surface conditioning [7]. Carbonization was performed using radio-frequency (RF)-assisted glow discharge cleaning (shortened to RG in the figure) with gas mixtures of $H_2 + CH_4$ or $D_2 + CD_4$. A movable RF antenna with a long stroke is inserted near to the center of the vacuum vessel and biased to a positive voltage, while the liner and limiters are at ground potential. The thickness of the carbon film produced by the glow discharge is 5–10 nm on stainless steel and last ~100 discharges in TEXTOR. The concentration of metal impurities is suppressed below 10^{-5} and the effective charge Z_{eff} drops close to unity. Carbonization expanded the operation regime for ohmically heated plasma in TEXTOR, as seen in the Hugill diagram in figure 1.4(b). A Hugill diagram plots the operation regime in a parameter space of the plasma safety factor and normalized electron density. The safety factor, q_{L}, is defined as $(RB_{\text{p}})/(aB_{\text{T}})$ and $B_{\text{p}} = I_{\text{p}}/(2\pi)$, where R and a are the major and minor radius, respectively, B_{T} and B_{p} are the toroidal and poloidal magnetic field strength, and I_{p} is the plasma current. Because the plasma confinement time is roughly proportional to the plasma current, a low q discharge has higher performance. However, the q_L values are usually limited to 2 due to magnetohydrodynamic instabilities near the plasma edge. Although the safety factor (q) was limited to 2.3 before carbonization, low q operation up to 2 was achieved after the carbonization. An expansion of the density limit by 50% was achieved after the walls' carbonized surface conditioning, as compared with the metallic walls before the carbonization. The density limit $n_{\text{e}}(0)R/B_{\text{T}}$ was increased

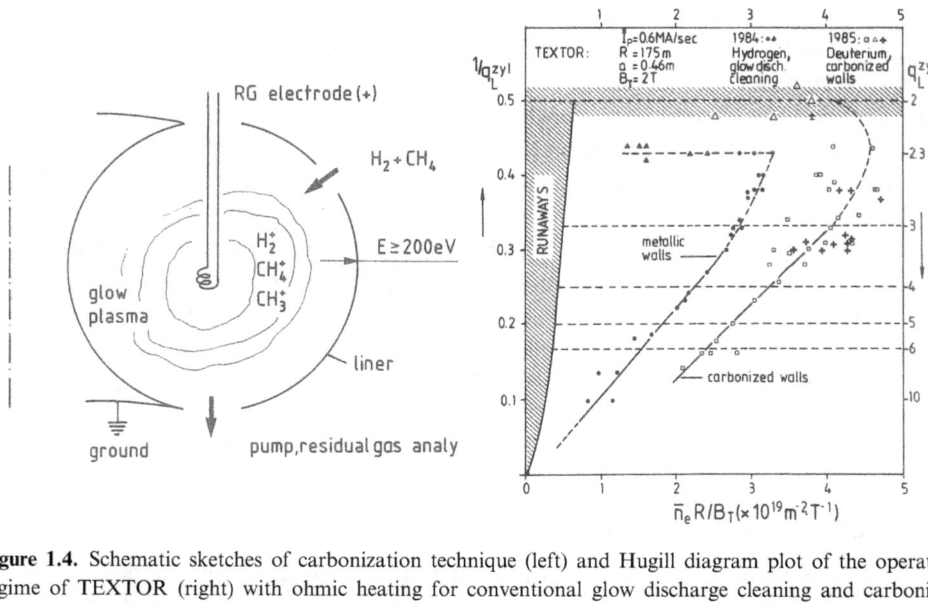

Figure 1.4. Schematic sketches of carbonization technique (left) and Hugill diagram plot of the operation regime of TEXTOR (right) with ohmic heating for conventional glow discharge cleaning and carbonized surface conditioning. Reprinted from [7], Copyright (1987), with permission from Elsevier.

by up to 5×19^{19} m^{-2} T^{-1} after carbonization. Thus carbonization has been found to be a useful tool to reduce radiation and expand operation regimes in ohmically heated plasma, and has been widely used.

Although carbonization reduces the concentration of high-Z impurities in the plasma, it does not have much impact on the reduction of oxygen impurities. In order to suppress this oxygen impurity influx, a boronization process has been applied in several devices. To this end, diborane (B_2H_6) was used as a mixture gas. Figure 1.5 shows spectroscopic data measured by visible spectroscopy and the intensity of the O VI line normalized to $\bar{n}_e$ as a function of $\bar{n}_e$ for various plasma currents with carbonization and at I_p = 340 kA for boronized walls and limiters [8, 9]. After boronization, in the Axially Symmetric Divertor Experiment (ASDEX), the iron-line radiation disappears from the spectra and the oxygen-line intensity is reduced by a factor of more than 5. As can be seen in figure 1.5(b), the O VI line intensity normalized by electron density increases, especially at a low plasma current, as the electron density increases after the carbonization. At the higher plasma current of 340 kA, the normalized O VI intensity starts to increase sharply at a higher electron density of 4×10^{13} m^{-3}. However, the O VI line intensity is small and shows only a small increase even at high electron density. From the line-averaged density of 3.5×10^{13} m^{-3}, the reduction of line intensity by boronization is larger than that by carbonization by a factor of 8.

Figure 1.6 shows the history of PFCs in various toroidal devices. As aforementioned, historically the material facing the plasma was metal prior to the mid-1980s. In the PLT, the hollow electron temperature profile was not observed after the replacement of the tungsten limiters with graphite limiters [2]. Although the replacement of the limiter materials with a low-Z material has a significant effect on the reduction of impurity accumulation in ohmic plasma, impurity accumulation still becomes an issue after auxiliary heating by NBI and ICRH is applied to increase the plasma temperature. This is due to the influx of high-Z impurities from the metal vacuum vessel wall. One approach to mitigate this is the technique of coating low-Z materials on the surface of the metal wall via glow discharge with low-Z materials

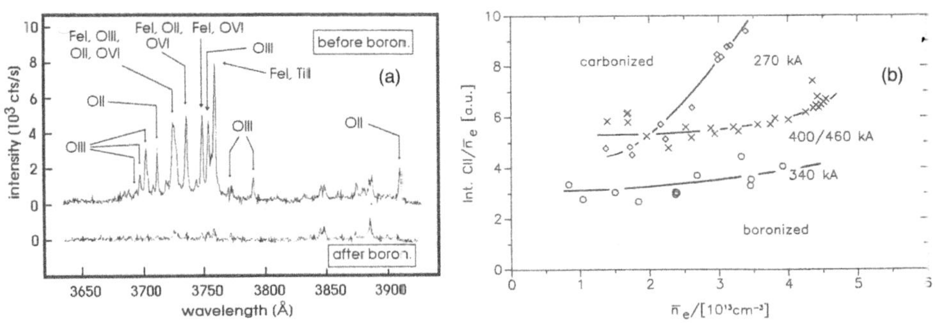

Figure 1.5. (a) Spectroscopic data measured by visible spectroscopy and (b) intensity of O vi line normalized to $\bar{n}_e$ as a function of $\bar{n}_e$ for various plasma currents with carbonization and at I_p = 340 kA for boronized walls and limiters. Panel (a) reprinted from [8], Copyright (1990), with permission from Elsevier. Panel (b) reprinted from [9], Copyright (1989), with permission from Elsevier.

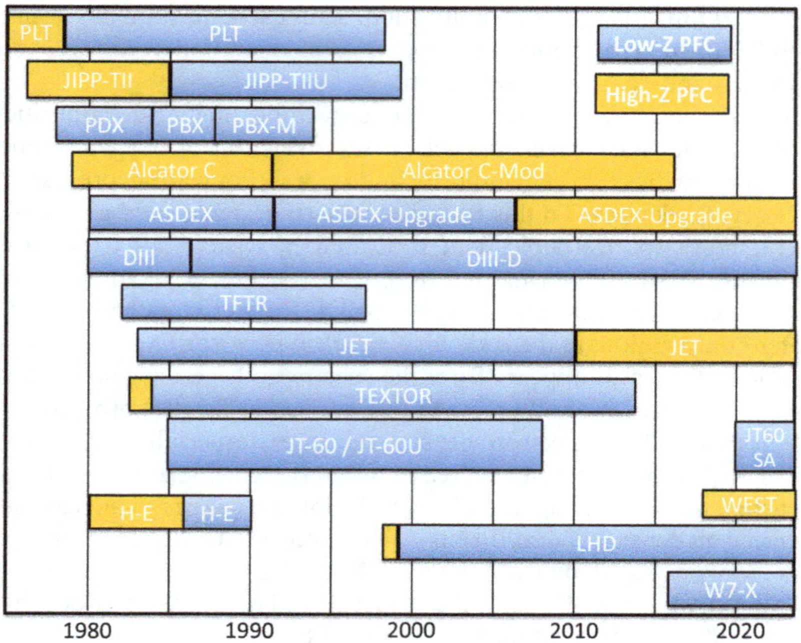

Figure 1.6. History of plasma-facing components (PFCs) in various toroidal devices.

such as boron and carbon. These techniques are known as boronization and carbonization, respectively, and have been widely applied in many devices (JITT-TIIU, TEXTOR, Heliotron E (H-E)) due to their convenience. The other approach is to install a carbon or beryllium divertor and tiles inside the vacuum vessel as PFCs. Titanium carbide (TiC)-coated molybdenum or TiC-coated Inconel were used in the early operation, in 1985, of the JAERI Torus-60 (JT-60) tokamak [10]. In order to avoid the melting of the divertor, all PFCs were replaced with carbon. In helical plasma, a slow cyclic oscillation between edge radiation collapse and recovery was observed during NBI-heated long-pulse discharges in early experiments with the Large Helical Device (LHD) using stainless steel divertor plates, which were referred to as 'radiation breathing'. This 'breathing' phenomenon was not observed after the installation of a graphite divertor [11].

The plasma performance in devices was thus improved by changing the PFCs from high-Z materials to low-Z materials in the 1980s (except for the example of Alcator C). Further, the application of carbon-based low-Z materials to PFCs has significantly improved plasma parameters in tokamak devices and helical plasma. However, there remain various serious problems, such as erosion and neutron damage, in the application of low-Z materials to future D-T burning machines [12]. It is also a major concern that these low-Z materials absorb hydrogen and its isotopes (deuterium and a tritium). The absorption of tritium to plasma-facing materials leads to the issue of high tritium retention inside the vacuum vessel in nuclear fusion devices such as the International Thermonuclear Experimental Reactor (ITER). Although tungsten has been proposed as a plasma-facing material for ITER, it is not clear whether high plasma performance can be obtained in the device with a metal divertor. It is also not known how to avoid the accumulation of high-Z (metal) ion impurities in plasma with improved confinement such as in H-mode and internal transport barrier plasma. As a result, experiments with high-Z PFCs (e.g., tungsten divertors) have seen a recent revival. The materials for the first wall in the ASDEX-Upgrade tokamak were modified to tungsten from graphite in 2002–2007. The Joint European Torus (JET) tokamak was modified to all-metal PFCs by installing a tungsten divertor in 2010. Another experiment in a tokamak with tungsten PFCs (the tungsten (W) Environment in Steady-state Tokamak (WEST)) started in 2018. It should be noted that boronization has also been frequently applied in the devices (Alcator C-Mod, ASDEX-Upgrade) with high-Z PFCs in order to achieve good plasma performance. The reduction of high-Z impurities in devices with high-Z PFCs remains a problem which has not been solved over the past 50 years.

1.3 Impurity transport

Impurity transport in magnetically confined plasmas has been an important issue in high-temperature plasmas confined with a strong magnetic field. This is because the partially ionized impurity ions cause an enhancement of radiation power loss from the plasma. When the radiation power is mainly emitted from the plasma core, it results in heating power loss in the plasma. However, when the radiation is mainly localized near the plasma periphery, the radiation loss contributes to the uniformity of the heat load to the vacuum vessel wall. Therefore, reduction of the impurity concentration is not always beneficial; rather, the control of the radial distribution determined by the impurity transport is more important than a simple reduction of impurity concentration. In contrast, low-Z impurities do not contribute to the power radiation but cause a dilution of bulk ions. Therefore, the concentration of impurities should be low enough, even for fully ionized impurity ions, that it does not contribute any radiation power loss.

If there are no impurities, the bulk ion transport is strongly coupled with the electron transport due to the neutralized condition of the plasma. However, the bulk ion density profile can differ from that of the electron density when there are some impurities in the plasma due to the decoupling between the electron particle transport and ion particle transport. Such decoupled ion particle transport would be important to predict the radial profiles of ions with different ion masses (e.g., hydrogen, deuterium, and tritium) in mixed-species plasma, which will be the case in nuclear fusion devices in the future. The assumption of the same ion transport for deuterium and tritium, which results in a constant isotope ratio in space, may not be valid. Impurity transport can be different from bulk ion transport in two important aspects. One is that the atomic mass and atomic charge of an impurity is different from that of bulk ions. The other is that the impurity transport is always decoupled from the electron transport because the electron density provided by a given impurity is relatively small. The importance of the second aspect is sometimes ignored. However, the non-linear characteristics of the impurity transport indicate its significance. Therefore, knowledge of the decoupled ion particle transport can be obtained by sophisticated impurity transport studies.

References

[1] Shimada M 1991 The present status of impurity transport investigations *Fusion Eng. Des.* **15** 325–41

[2] Meservey E B, Arunasalam V, Barnes C, Bol K, Bretz N and Cohen S *et al* 1980 The effect of plasma surface interactions on plt plasma parameters *J. Nucl. Mater.* **93–94** 267–71

[3] Hawryluk R J, Bol K, Bretz N, Dimock D, Eames D and Hinnov E *et al* 1979 The effect of current profile evolution on plasma-limiter interaction and the energy confinement time *Nucl. Fusion* **19** 1307–17

[4] Isler R C, Kasai S, Murray L E, Saltmarsh M and Murakami M 1981 Impurity sources and accumulation in ohmically heated ISX-B discharges *Phys. Rev. Lett.* **47** 333–7

[5] Isler R C, Murray L E, Kasai S, Arnurius D E, Bates S C and Crume E C *et al* 1981 Influence of neutral-beam injection on impurity transport in the ISX-B tokamak *Phys. Rev. Lett.* **47** 649–52

[6] Ida K, Fonck R J, Sesnic S, Hulse R A, LeBlanc B and Paul S F 1989 Impurity behaviour in PBX L- and H-mode plasmas *Nucl. Fusion* **29** 231–50

[7] Winter J 1987 Carbonization in tokamaks *J. Nucl. Mater.* **145-147** 131–44

[8] The ASDEX TeamThe ICRH TeamThe LH TeamThe NI TeamSchneider U and Poschenrieder W *et al* 1990 Boronization of ASDEX *J. Nucl. Mater.* **176-177** 350–6

[9] Winter J, Esser H G, Konen L, Philipps V, Reimer H and v Seggern J *et al* 1989 Boronization in textor *J. Nucl. Mater.* **162-164** 713–23

[10] Takatsu H, Ando T, Yamamoto M, Arai T, Kodama K and Ohkubo M *et al* 1988 Present knowledge about the materials behavior in JT-60 *J. Nucl. Mater.* **155-157** 27–40

[11] Peterson B J, Nakamura Y, Yamazaki K, Noda N, Rice J and Takeiri Y *et al* 2001 Role of core radiation during slow oscillations in LHD *Nucl. Fusion* **41** 519–25

[12] Tanabe T, Noda N and Nakamura H 1992 Review of high Z materials for PSI applications *J. Nucl. Mater.* **196-198** 11–27

Katsumi Ida and Naoki Tamura

Chapter 2

Transport model

The impurity transport model is summarized in this chapter. In this, atomic processes such as ionization and recombination are important, because such atomic processes determine the charge state of the impurities and the level of radiation loss. Inside the last closed-flux surface, the impurity transport is governed by neoclassical and turbulent transport. As neoclassical transport effects, the inward pinch and temperature screening are described. Ripple transport and poloidal variation of the electrostatic potential, which become important in non-axisymmetric plasma (i.e., helical plasma) are discussed, Turbulence-driven flux, which has a significant impact on impurity transport both in axisymmetric (tokamak) and non-axisymmetric (helical) plasmas, is described. Finally, transport in the open-flux surface region, where transport parallel to the magnetic field plays an important role, is discussed.

2.1 Atomic processes

The transport of impurities is more complicated than the particle transport because the impurity density for a given charge state, n_z, is determined by the transport and atomic processes as

$$\frac{\partial n_z}{\partial t} = -\frac{1}{r}\frac{\partial}{\partial r}(r\Gamma_z) + S_{z-1}n_{z-1} - (S_z + R_z)n_z + R_{z+1}n_{z+1} - \frac{n_z}{\tau_{||}} + F_{z-1}. \quad (2.1)$$

Here, S_z and R_z are the total ionization rate and the total recombination rate, including dielectronic, radiative, and charge-exchange recombination, for the charge state Z. $n_z/\tau_{||}$ is the loss term for the plasma flow parallel to the magnetic field line in the region of the magnetic field, such as in the scrape-off layer (SOL) and stochastic magnetic field region. $\tau_{||}$ is the time for parallel loss. F_z is the term for the volume source for neutral (i.e., $F_z = 0$ for $z \neq 0$). The radial flux of the impurity density, Γ_z, is given by a simple diffusion convection model as

doi:10.1088/978-0-7503-1451-0ch2

$$\Gamma_z(r) = -D_z\frac{\partial n_z}{\partial r} - V_z^{\text{conv}}n_z. \tag{2.2}$$

Here, D_z is the diffusion coefficient and V_z^{conv} is the convection velocity. When the diffusion coefficient is assumed to be constant in space and independent from the charge state, and when the convection velocity is proportional to the radius, D_z and V_z^{conv} can be expressed using a peaking factor, $c_{v,z}$, as

$$D_z(r) = D, \tag{2.3}$$

$$V_z^{\text{conv}}(r) = -c_{v,z}\frac{2Dr}{a^2}, \tag{2.4}$$

where a is the minor radius of the plasma.

By using the peaking factor $c_{v,z}$, the total impurity density can be given by the following expression:

$$\sum_z n_z(r) \propto \exp\left(\frac{-c_{v,z}r^2}{a^2}\right). \tag{2.5}$$

The peaking factor $c_{v,z}$ is often used to express the magnitude and sign of the convection term in a simple transport model. A positive peaking factor represents inward convection (impurity accumulation due to particle pinch) of the impurity, while a negative peaking factor represents outward convection, which results in hollow impurity density profiles.

In the plasma core there is no plasma flow parallel to the magnetic field line nor source for neutral. Therefore, the impurity transport in a steady-state condition ($\partial n_z/\partial t = 0$) can be expressed as

$$\frac{1}{r}\frac{\partial}{\partial r}(r\Gamma_z) = S_{z-1}n_{z-1} - (S_z + R_z)n_z + R_{z+1}n_{z+1}. \tag{2.6}$$

Here, the left-hand side of the equation is the transport term, while the right-hand side of the equation gives the atomic process terms. Figure 2.1 shows the rate coefficients of electron impact ionization and total recombination of (a) carbon C^{5+} and C^{6+} and (b) tungsten W^{29+} and W^{30+} given by ADF11 data (unresolved) from the OPEN-ADAS system[1]. In the ADF11 data (unresolved), all the coefficients depend on the free electron temperature and density and are produced by collisional-radiative models. Data at two levels of refinement are present, namely, unresolved, in which only the ground states of ions are assumed to be dominant species. As the plasma temperature is raised, the ionization rate sharply increases, while the recombination rate drops. The electron temperature at the cross point, where the recombination rate is equal to the ionization rate, is ~0.1 keV for C^{5+}–C^{6+}, while it is 1.5 keV for W^{29+}–W^{30+}. Although the ionization rate coefficient is not dependent on the electron density, the recombination rate coefficient increases as the electron

[1] Available online: http://open.adas.ac.uk.

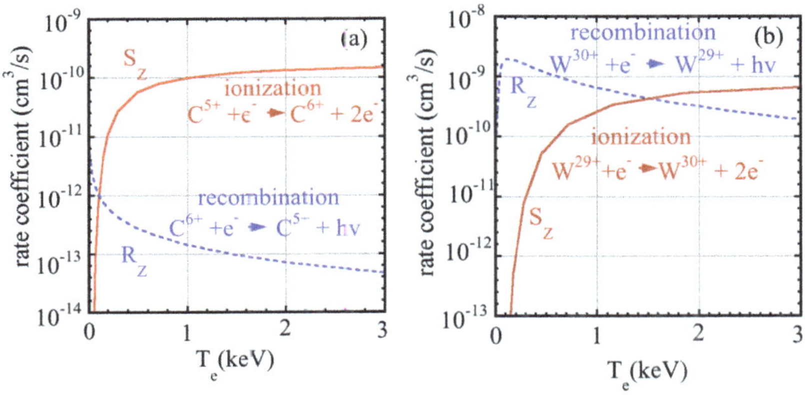

Figure 2.1. Rate coefficients of electron impact ionization and total recombination of (a) carbon C^{5+} and C^{6+}, and (b) tungsten W^{29+} and W^{30+} for an electron density of 1×10^{19} m^{-3}, given by ADF11 of OPEN-ADAS.

density is increased. Here, the recombination rate at an electron density of 1×10^{19} m^{-3} is plotted. The low-Z carbon impurity is fully ionized in the plasma core, but the high-Z tungsten impurity is only partially ionized in the plasma core. Therefore, the low-Z impurity does not cause radiation loss, whereas the high-Z impurity causes radiation loss in the plasma core or plasma center when impurity accumulation occurs. For the low-Z impurity (carbon), the ionization rate is much larger than the recombination rate for the high-temperature toroidal plasma, and the recombination term can be neglected as

$$\frac{1}{r}\frac{\partial}{\partial r}(r\Gamma_z) = S_{z-1}n_{z-1} - S_z n_z. \tag{2.7}$$

Then the radial flux of the transport term is balanced to the source term due to the ionization for low-Z impurity. In contrast, the ionization–recombination balance is mainly determined by the atomic process terms for high-Z impurity. The ionization–recombination balance is significantly modified by the transport process. As the diffusion is increased, the ionization–recombination balance shifts to a low ionization state. Therefore, the diffusion coefficient can be estimated from the ionization state if the electron temperature is known.

Therefore, when the outward radial flux Γ_z is reduced by a small diffusion coefficient or inward convection, the reduction of the outward radial flux contributes to an increase of n_z and decrease of n_{z-1}, and a higher charge state becomes more dominant. The charge state distribution in toroidal plasmas is not determined by atomic processes alone and is strongly influenced by radial transport in the plasma. Figure 2.2 shows the influence of this transport on the charge state distribution. Two different models for the radial profile of the diffusion coefficient, $D(\rho)$, are introduced to investigate the transport effect on the charge state distribution of the impurities, while the convection velocity is set to be zero for simplicity. Here, ρ is the normalized minor radius, defined as r/a, and set to be 0 at the magnetic axis and 1 at the plasma edge. When the core diffusion coefficient is small ($D(0.42) = 0.01$ m^2 s^{-1}),

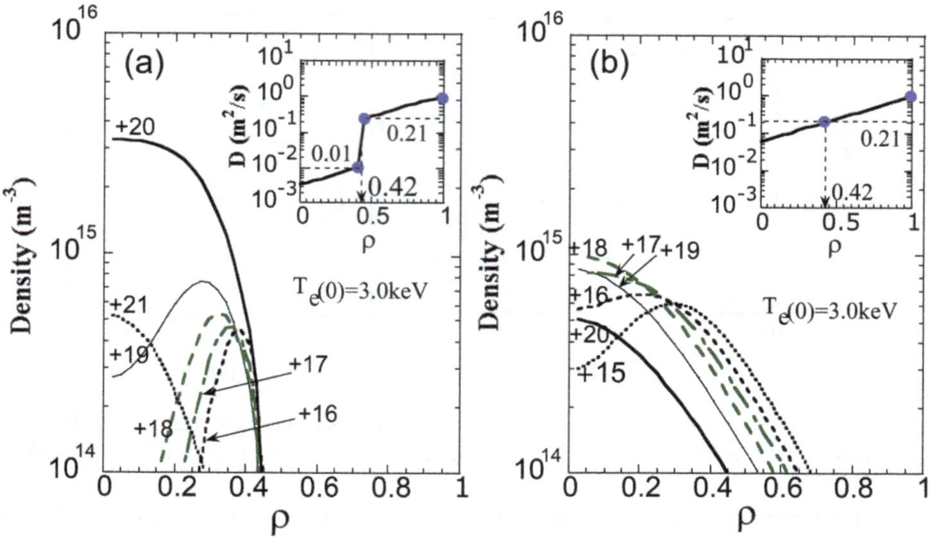

Figure 2.2. Radial profiles of titanium (Ti) density of each charge state calculated using equation (3) for plasma in the CHS with a central electron temperature of 3 keV for (a) $D(0.42) = 0.01\ \mathrm{m^2\ s^{-1}}$ and $V^{\mathrm{conv}} = 0$, and (b) $D(0.42) = 0.21\ \mathrm{m^2\ s^{-1}}$ and $V^{\mathrm{conv}} = 0$. Reprinted from [1], with the permission of AIP Publishing.

the helium-like titanium impurity (Ti^{20+}) is dominant in the core region ($\rho < 0.42$), while the beryllium-like and boron-like titanium impurities (Ti^{18+} and Ti^{17+}) become dominant when the core diffusion coefficient becomes large ($D(0.42) = 0.21\ \mathrm{m^2\ s^{-1}}$) in Compact Helical System (CHS) plasma [1]. The radial profile of the electron temperature and the diffusion coefficient near the plasma edge is identical for these two cases. The small diffusion coefficient in the core region ($\rho < 0.42$) increases the higher charge state impurity as well as the peaking of the radial profile of the impurity density.

The influence of radial transport on the charge state distribution differs significantly between the inside and outside the core regions with low diffusivity. Although the radial flux due to the transport decreases the density of Ti^{20+} inside the core region, it actually increases the density of Ti^{17+}–Ti^{19+} outside the core region at $\rho = 0.5$–0.6. This is because once the impurity is ionized to a higher charge state in the core, where the temperature is high, it travels to the edge region, where the temperature is lower, for the charge state before it is recombined. The charge state of the impurity is not determined by the local temperature but is determined by the highest temperature at the magnetic axis due to this transport effect. This transport effect is quite beneficial in charge-exchange spectroscopy, where the charge-exchange reaction between fully ionized impurity ions and the neutral beam is utilized. Because of this transport effect, charge-exchange spectroscopy can be utilized even at the plasma edge, where the local temperature is not high enough to produce fully ionized impurities. In fact, the ion temperature and plasma flow can be measured with fully ionized carbon at the plasma edge, where the electron temperature is only 100 eV and much lower than the ionization potential of hydrogen-like carbon of 490 eV.

2.2 Neoclassical and turbulent transport in the closed-flux surface (core) region

In the particle, momentum, and heat transport of bulk ions, the contribution of neoclassical transport is small and is usually neglected because the contributions of turbulent transport (diffusion, perpendicular viscosity, and thermal diffusivity) are much larger than the contributions of collisional transport (that is, neoclassical transport). However, the number of impurity particles is much smaller than the number of bulk ions. As such, the transport of impurities is mainly determined through the collision process between impurity and bulk ions rather than the collision process between impurities. Therefore, the collision process and neoclassical transport become much more important in impurity transport, which differs from other transport in the plasma [2–4].

The neoclassical radial flux of the impurity is given by

$$\Gamma_z^{\mathrm{NC}} = -D_z^{\mathrm{NC}} \frac{\partial n_z}{\partial r} + \Gamma_z^{\mathrm{friction}} + \Gamma_z^{\mathrm{ripple}}, \tag{2.3}$$

where, in the following equations, D_z^{NC} is the neoclassical diffusion coefficient for a species of impurity of charge number Z on its own density gradient, n_z is the impurity density, n_j is the density of the plasma species j, T_{i} is the ion temperature, q_z is the electric charge of the impurity species of charge Z, and E_r is the radial electric field. The $g_{n_{j\to z}}$, $g_{T_{\mathrm{i}}}$ are the coefficients for each term and depend on the magnetic field and collisionality for each species. The first term is a diffusion term due to the collision between impurities, and the second is the convection terms, which are independent from the gradient of the impurity density. The first and second terms of the convection terms are the guiding-center drift of the impurity in the radial direction due to collisions between bulk ions and impurities, which are referred to as friction terms. The third term of the convection terms is the ripple transport, which can be significant in non-axisymmetric toroidal plasma. However, this term vanishes in plasma with toroidal symmetry, where the collisional transport is independent from the radial electric field ($\Gamma_z^{\mathrm{ripple}} = 0$ for a tokamak).

2.2.1 Inward pinch and temperature screening

$$\Gamma_z^{\mathrm{friction}} = D_z^{\mathrm{NC}} n_z \left[\sum_{j \neq z} g_{n_{j\to z}} \frac{\partial n_j}{n_j \partial r} + g_{T_{\mathrm{i}}} \frac{\partial T_{\mathrm{i}}}{T_{\mathrm{i}} \partial r} + g_{T_{\mathrm{e}}} \frac{\partial T_{\mathrm{e}}}{T_{\mathrm{e}} \partial r} \right]. \tag{2.9}$$

In the collision process between impurity and bulk ions, the guiding-center drift of the impurity in the radial direction is determined by the density gradient and the temperature of the bulk ions. In a plasma with a peaked density profile, the collisions between impurity and bulk ions are larger in the inner region than in the outer

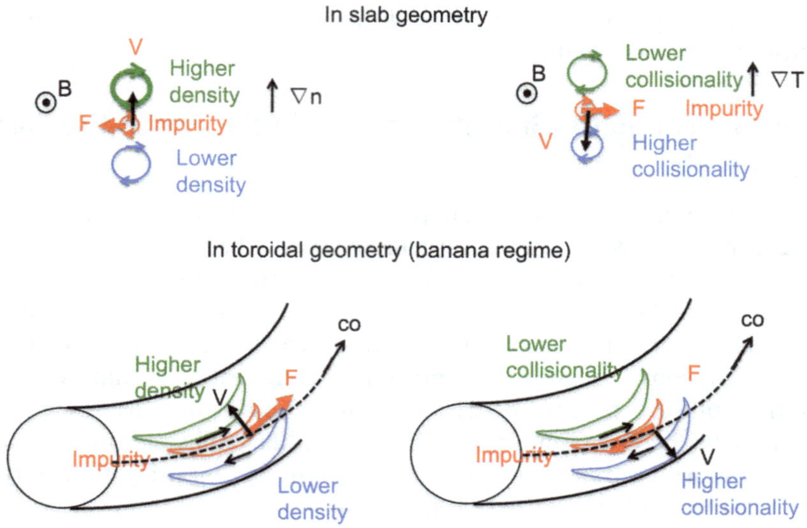

Figure 2.3. Schematics outlining principle of the guiding-center drift of the impurity in the radial direction in a slab geometry (top) and in a banana regime in toroidal geometry (bottom).

region, and the radial drift of the impurity due to the frictional force points inward, as seen in figure 2.3. This inward drift due to the density gradient is referred to as the 'inward pinch', and has been recognized to be a severe problem in the accummulation of impurities, because this process causes the impurity density gradient to become larger than the bulk ion gradient by a power of charge z and results in much more peaked impurity profiles than the bulk ions in a steady state, with the following relations [5]:

$$\frac{1}{n_z}\frac{\partial n_z}{\partial r} = \frac{q_z}{n_i}\frac{\partial n_i}{\partial r}. \tag{2.10}$$

In contrast, in a plasma with a peaked temperature profile, the collision frequency between impurity and bulk ions in the inner region is smaller than in the outer region, and the radial drift of the impurity due to this frictional force points outward. This outward drift of the impurity is known as the 'temperature screening' effect. In a plasma with a peaked density and temperature profile, there are two drifts: One the inward impurity drift due to the density gradient, and the other the outward impurity drift due to the temperature gradient of the bulk ions. In this case, the impurity density profile is determined by the balance between the inward pinch and temperature screening effect. In a banana regime, that is, particles in a toroidal geometry, the inward pinch due to the density gradient and temperature screening owing to the temperature gradient co-exist in the plasma, which has both a peaked density and temperature profile.

A negative density gradient causes the inward (smaller minor radius, r) radial drift of impurities, and a negative temperature gradient causes the outward (larger minor radius, r) radial drift of impurities through the frictional force between bulk ions and impurities in a toroidal geometry, as illustrated in figure 2.4. In a tokamak, the magnitude of the toroidal magnetic field is roughly proportional to the major radius, R, and the gradient of the magnetic field strength causes the downward (negative z-direction) drift of ions and upward (positive z-direction) drift of electrons, resulting in up–down charge separation. However, this charge separation is canceled by the pitch of the magnetic field (e.g., poloidal magnetic field). These neoclassical effects are also clearly demonstrated by kinetic calculation. Figures 2.4(b) and (c) show distributions of impurity position projected to a poloidal cross-section calculated with kinetic modeling [6]. Here, the blue circle is the initial position of the impurity and the red points are the positions scattered by neoclassical collisions. The distribution of the impurity position becomes a wide ring with a width 0.2–0.5 m due to the diffusion process. The convection of impurity transport results in inward and the outward shifts of the distribution, as seen in figures 2.4(b) and (c).

In a toroidal geometry, the plasma flow parallel to the magnetic field is produced by the incompressible condition. This flow is known as a Pfirsch–Schlüter (PS) flow. The projection of this PS flow on poloidal cross-sections is also indicated by arrows in figures 2.4(b) and (c). The inward shift by density gradient and outward drift by temperature gradient are clearly reproduced by the calculation. In the case in which the ion density gradient is small in a plasma with a flat density profile and peaked ion temperature profile, the outward impurity flux driven by the ion temperature gradient overcomes the inward pinch and is even balanced by the inward impurity diffusion flux due to the positive impurity density gradient, which results in a hollow impurity density profile [7].

2.2.2 Ripple transport

$$\Gamma_z^{\text{ripple}} = D_z^{\text{NC}} n_z \left[\alpha_z^{12} \frac{\partial T_\text{i}}{T_\text{i} \partial r} - \frac{q_z e E_\text{r}}{T_\text{i}} \right]. \tag{2.11}$$

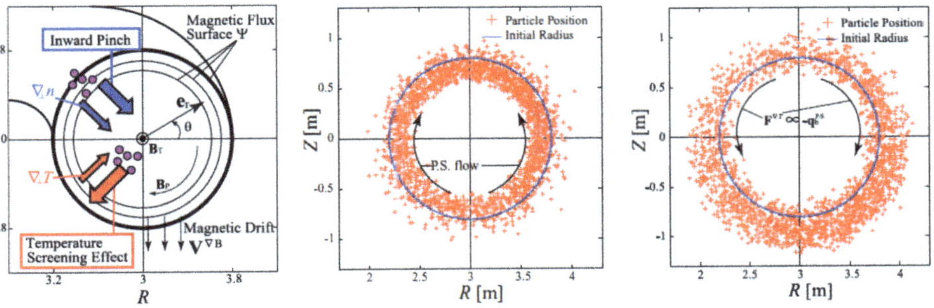

Figure 2.4. (a) Schematic of neoclassical inward pinch and temperature screening effect, and distribution of impurity position projected to a poloidal cross-section calculated with kinetic modeling for the reference cases of (b) neoclassical inward pinch and (c) neoclassical temperature screening effect. Reproduced from [6].

In a plasma with a non-axisymmetric configuration, such as in a stellarator or heliotron, ripple transport becomes important. This ripple transport is sensitive to the radial electric field, which is determined by the ambipolar condition of the electron and ion flux of bulk ions, $\Gamma_e^{NC}(E_r) = \Gamma_i^{NC}(E_r)$, because both the electron and the ion flux driven by turbulence are nearly ambipolar.

By using neoclassical theory, the third term of the convection terms driven by the radial electric field can be expressed as

$$eE_r = \frac{T_e T_i}{T_e D_i + T_i D_e}\left[(D_i - D_e)\frac{\partial n_i}{n_i \partial r} + D_i \delta_i^{12}\frac{\partial T_i}{T_i \partial r} - D_e \delta_e^{12}\frac{\partial T_e}{T_e \partial r}\right], \tag{2.12}$$

where D_i and D_e are the diffusion coefficient of ions and electrons, respectively, and

$$\delta_{i,e}^{12} = \frac{L_{i,e}^{12}}{D_{i,e}} - \frac{3}{2}. \tag{2.13}$$

The radial electric field is roughly proportional to the electron pressure gradient, and becomes positive for the electron root ($D_e \gg D_i$). In contrast, for the ion root ($D_i \gg D_e$) the radial electric field is roughly proportional to the ion pressure gradient, and becomes negative. A positive radial electric field results in convection outward (from the center to the edge of the plasma), while a negative electric field results in convection inward (from the edge to the center of the plasma). Then, the impurities usually accumulate in the center in ion-root plasma, while no impurity accumulation is expected in electron-root plasma. Figure 2.5 shows simulation results for a fully ionized argon density profile after an argon gas puff without recycling for two plasma with different collisionalities. One is the case for a higher density and higher collisionality, where the ion root is expected in the whole plasma region, and the other is for a lower density and lower collisionality, where the electron root is expected to appear in some plasma regions. In the case of higher density, the argon impurity accumulates to the plasma center ($\rho = 0$) and the central argon density keeps increasing in time. In contrast, the central argon impurity reaches a maximum at 0.35 s after the argon puff and decays in time afterwards, and

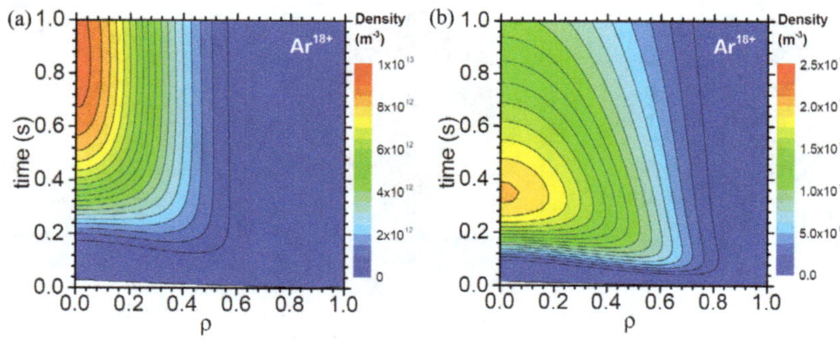

Figure 2.5. Temporal evolutions of the spatial distribution of the argon density of Ar^{18+} in (a) a high-density plasma and (b) a low-density plasma. Reproduced from [10].

the central argon density decreases by a factor of 2 at 1.0 s after the argon puff. Impurity transport in a helical plasma is very sensitive to collisionality and the radial electric field due to ripple transport, which is discussed in chapter 5.1. In neoclassical transport, both the frictional term (comprising the inward pinch and temperature screening effect) and ripple transport term are determined by the density and temperature gradients of the bulk ions. The radial flux owing to this ripple transport is significantly reduced in the so-called optimized helical configuration, where the effect of ripple is minimized. Therefore, radial flux due to the friction force, which has otherwise been neglected in helical plasmas, has come to be recognized as important [8]. In fact, the simultaneous peaking of bulk ions and hollowing impurity profile observed in helical plasma with an internal transport barrier in the Large Helical Device (LHD) suggests the coexistence of the inward radial flux of bulk ions and the outward radial flux of impurities due to the friction force [9].

2.2.3 Poloidal variation of electrostatic potential

Although the radial electric field is not constant on a magnetic flux surface, the electrostatic potential is nearly constant on a magnetic flux. Therefore, the poloidal electric field is much smaller than the radial electric field and vanishes in time averaging. However, when both the density and electrostatic potential have variation on the magnetic flux, the density n and electrostatic potential Φ can be expressed as $n = n_0(r) + n_1(r, \theta, \phi)$ and $\Phi = \Phi_0(r) + \Phi_1(r, \theta, \phi)$. Poloidal variation of the electrostatic potential produces a poloidal electric field and radial flux due to $E_\theta \times B_\phi$. The radial flux of the impurity due to variation of the electrostatic potential can be expressed as

$$\Gamma_{nz} = \langle n_z v_r \rangle_{\mathrm{flux}} = \left\langle n_z \frac{B \times \nabla\Phi_1}{B^2} \right\rangle_{\mathrm{flux}}, \tag{2.14}$$

where $\langle\rangle_{\mathrm{flux}}$ denotes the flux average. This is a steady-state radial flux driven by the electrostatic potential $\Phi_1(r, \theta, \phi)$, which is constant in time. This poloidal asymmetry of the electrostatic potential is due to the three-dimensional (3D) effect of the magnetic field, which can be important in helical plasmas [11–14]. In an axisymmetric system such as a tokamak, the poloidal asymmetry of the electrostatic potential disappears. Figure 2.6 illustrates how the radial flux of the impurity is driven by the poloidal asymmetry of the potential and density variation in helical plasmas. In this case, the space potential has up–down asymmetry (higher potential upward) and the impurity density has in–out asymmetry (higher density in large major radius). The up–down asymmetry of the space potential produces the steady-state poloidal electric field, E_θ, and a radial velocity of $E_\theta \times B$. This $E_\theta \times B$ velocity is outward in the low-field side (larger major radius) and inward in the high-field side (smaller major radius). Since the impurity density is larger in the low-field side than in the high-field side, the net radial flux of the impurity (the product of the radial velocity and density) can be outward by flux averaging.

The poloidal asymmetry of the space potential is not always up–down asymmetric and varies along the toroidal direction. This complicated structure of the perturbed

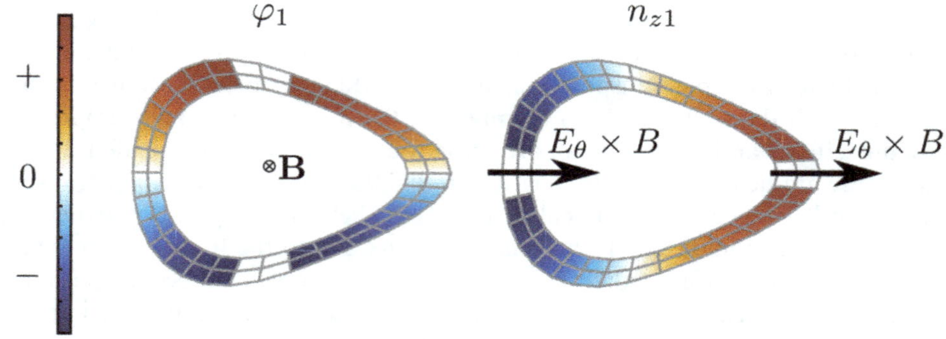

Figure 2.6. Potential φ_1 and impurity density variation n_{z1} on a magnetic flux in a helical plasma. Courtesy of J. A. Alonso.

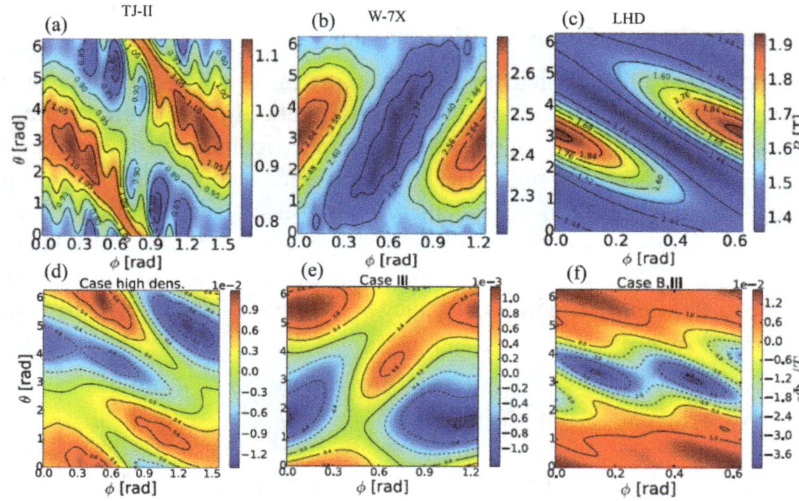

Figure 2.7. Contours of (a)–(c) magnetic field strength and (d)–(f) perturbation of space potential in (a) and (d) TJ-II, (b) and (e) W-7X, and (c) and (f) LHD at a magnetic flux surface of $\rho = 0.6$. Here, θ is the poloidal angle and ϕ is the toroidal angle. Reproduced from [13].

space potential, $\Phi_1(r, \theta, \phi)$, depends on the 3D magnetic structure. Figure 2.7 shows contours of the magnetic field strength and perturbation of the space potential in the Tokamak de la Junta II (TJ-II), Wendelstein 7-X (W7-X), and LHD devices. Here, $\theta = 0$ and π stands for the horizontal direction (0 outward, and π inward), while $\theta = \pi/2$ and $3\pi/2$ stands for the vertical direction, ($\pi/2$ upward, and $3\pi/2$ downward). In the case of the LHD, $\phi = 0$ stands for the vertically elongated cross-section, while $\phi = \pi/10$ stands for the horizontally elongated cross-section. The poloidal asymmetry of the space potential in the LHD is mainly an in–out asymmetry, while the poloidal asymmetry of the space potential in W7-X is an up–down asymmetry except for the cross-section of $\phi = \pi/5$. The poloidal asymmetry of the space potential in W7-X and

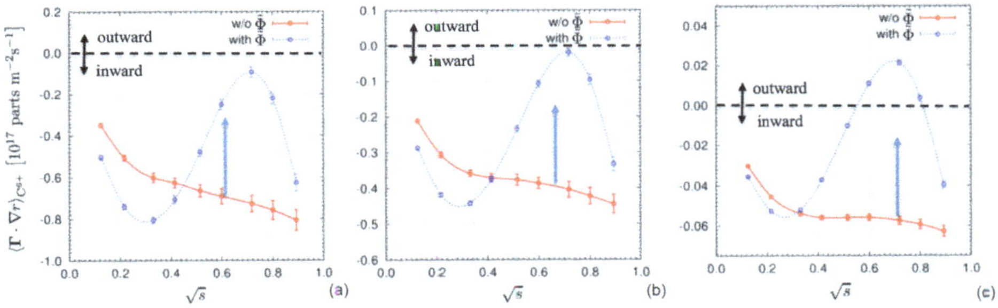

Figure 2.8. Radial profiles of the particle flux density of carbon (C^{6+}), neon (Ne^{8+}), and iron (Fe^{20+}) in ion-root plasma with negative radial electric field in the LHD. Reproduced from [11].

TJ-II varies between in–out and up–down depending on the toroidal angle. The magnitude of the perturbation of the space potential is on the order of 1% in the LHD and TJ-II, which are much larger than that in W7-X (0.1%). Therefore, modification of the impurity transport due to poloidal variation of the electrostatic potential can be important in the LHD and TJ-II, but not in W7-X.

Figure 2.8 shows radial profiles of the particle flux density of carbon (C^{6+}), neon (Ne^{8+}), and iron (Fe^{20+}) with (blue) and without (red) electrostatic potential variation on the flux surface [11] in the LHD. Here, a positive particle flux represents inward (toward the plasma center) flux, while a positive particle flux represents outward (toward the plasma edge) flux, and the x-axis is the normalized minor radius, defined by $\sqrt{s}$. An impurity particle flux without electrostatic potential variation would be inward and cause impurity accumulation. Such impurity accumulation would be a serious problem in a long pulse discharge because the central radiation results in degradation of the plasma performance. However, this inward flux becomes weak, especially in the outer region of the plasma ($\sqrt{s} > 0.5$), when the electrostatic potential variation is taken into account. For the high-Z impurity of Fe^{20+}, a region of positive particle flux (outward flux) appears in the plasma ($0.55 < \sqrt{s} < 0.8$), which could play the role of a barrier to prevent impurity accumulation in the core.

2.2.4 Fluctuation-driven flux

A fluctuating electrostatic potential always exists in both asymmetric and non-asymmetric toroidal plasmas, and its contribution to the radial flux of the impurity is comparable to the radial flux due to friction force in tokamak devices and ripple transport in helical systems. The turbulence-driven impurity flux is written as $\langle\langle \tilde{n}_z \tilde{v}_r \rangle_{\text{time}} \rangle_{\text{flux}}$, where $\langle\rangle_{\text{time}}$ is the time average [15, 16]. The radial flux of the impurity density driven by turbulence is expressed as [17]

$$\Gamma_{nz} = \langle\langle \delta n_z v_{E\times B} \rangle_{\text{time}} \rangle_{\text{flux}} = -n_z \rho_s c_s \left\langle \left\langle \tilde{n}_z \frac{1}{r} \frac{\partial \tilde{\phi}}{\partial \theta} \right\rangle_{\text{time}} \right\rangle_{\text{flux}}. \tag{2.15}$$

Here, the angled brackets $\langle\langle\rangle\rangle$ stand for a time and space average over all unstable modes on the magnetic flux surface. The sign of a turbulence-driven impurity flux is determined by the phase relation between the density fluctuation ($\tilde{n}_z$) and potential fluctuation ($\tilde{\phi}$). The radial flux of the impurity can be inward or outward depending on the type of fluctuations, such as the ion temperature gradient (ITG) mode, electron temperature gradient (ETG) mode, and trapped electron (TE) mode.

Figure 2.9(a) shows a fluctuation of the electric potential $\tilde{\phi}$ on a cross-section of a flux tube calculated using the non-linear gyrokinetic simulation code GENE[2] for the TE mode. The spatial resolution for each flux tube is 125 × 125 main ion gyro radii, where x is the radial direction and y is the poloidal direction, respectively. The size of the fluctuation of the electric potential is 10–20 times the ion gyro radii. As seen in figures 2.9(b) and (c), a positive peaking factor, PF($=-RV_z/D_z$), is expected both for the ITG mode and the TE mode. Here, the PF is defined as the ratio of convective velocity and diffusion, and a positive PF (negative V_z) represents inward convection of the impurity transport. TE mode and ITG mode turbulence have different Z dependences of PF. In the case of the TE mode, the PF has a tendency to decrease as the Z values increase, which suggests that TE mode turbulence does not cause high-Z impurity accumulation. In contrast, in ITG mode plasma, the PF has a tendency to increase as the Z values increase, which suggests that the accumulation of high-Z impurities could be a problem. The low PF at the lowest Z of 2 indicates that helium ash removal would not be a problem even for an ITG-dominant plasma. The difference in the fluctuating mode between helical and tokamak plasmas is one of the

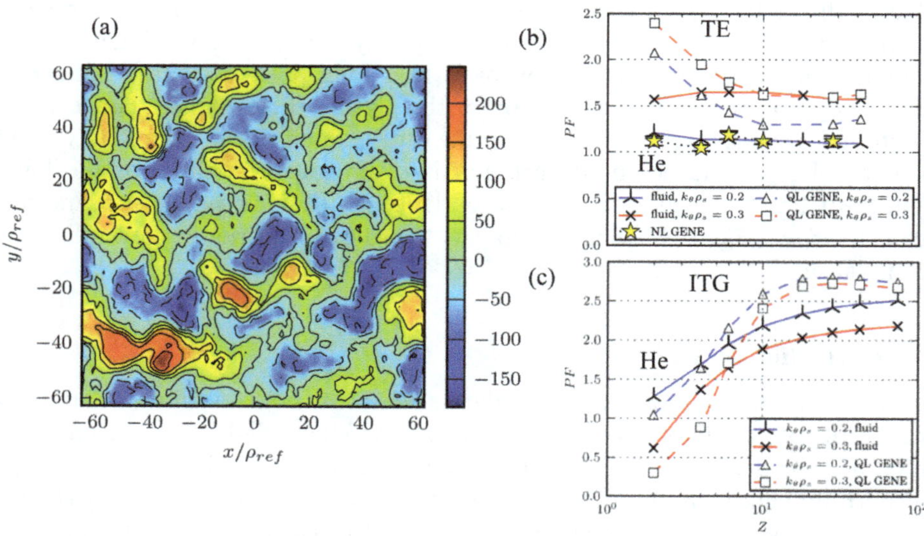

Figure 2.9. (a) Fluctuation of the electric potential $\tilde{\phi}$ on a cross-section of a flux tube for the trapped electron mode, and Z dependences of the peaking factor (PF) of impurity density for (b) trapped electron (TE) mode and (c) ion temperature gradient (ITG) mode. Reprinted from [17], with the permission of AIP Publishing.

[2] Available online: http://www.genecode.org/.

candidates for the mechanism causing the different turbulence-driven behavior of impurities in these devices [18, 19].

2.3 Transport in the open-flux surface region

Impurity transport in the open-flux surface is mainly determined by the transport parallel to the magnetic field. In this case, the profile of the impurity density is determined by the competing friction force and ion thermal force acting on the impurity. In a fluid model, the impurity transport along the magnetic field line is expressed by the following classical force balance [20]:

$$m_z \frac{\partial V_{z\|}}{\partial t} = \underbrace{-\frac{1}{n_z}\frac{\partial T_z n_z}{\partial s}}_{\text{pressure gradient}} + \underbrace{ZeE_{\|}}_{\text{electrostatic}} + \underbrace{m_z \frac{V_{i\|} - V_{Z\|}^{\text{imp}}}{\tau_s}}_{\text{friction}} \\ \underbrace{+ 0.71 Z^2 \frac{\partial T_e}{\partial s} + 2.6 Z^2 \frac{\partial T_i}{\partial s}}_{\text{temperature gradient}}, \tag{2.16}$$

where m_z and $V_{z\|}$ are the mass and parallel flow velocity of an impurity with an ion charge Z, T_z and n_z are the temperature and density of the impurity, $E_{i\|}$ is the electric field parallel to the magnetic field, $V_{i\|}$ is the parallel flow of the bulk ions, T_e and T_i are the electron and ion temperature of the bulk ions, and s is the coordinate along the magnetic field.

As seen in figure 2.10, the flow of impurities along the magnetic field is determined by the balance between the electrostatic force, the friction force, and the temperature gradient force (thermal force). The first two terms on the right-hand side express the pressure gradient of the impurity and the electrostatic force working on the impurity, while the last three terms originate in the collision with the background plasma. The third term on the right-hand side is the friction force exerted by the background ions moving with fluid velocity, V_i, where τ_s is the collision time between the impurity and background plasma. This friction force drags the impurity toward the plasma flow direction (for example, toward the divertor region, assuming that the plasma flow is directed toward divertor plates).

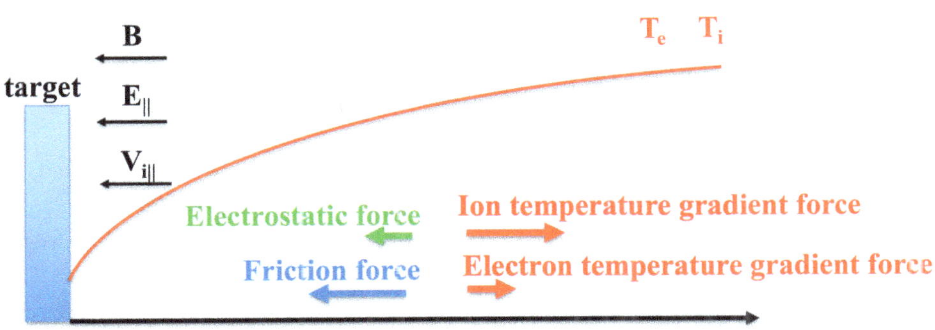

Figure 2.10. The principal parallel forces acting on impurity ions in the SOL.

The fourth and the fifth terms express the temperature gradient force of electrons and ions, respectively, which appears with the parallel temperature gradients and has its origin in the temperature dependence of the Coulomb collision, $\nu \propto T^{-3/2}$. This term is also called the thermal force [21] and drives the impurity upstream because the temperature gradient parallel to the magnetic field points upstream and toward higher temperatures. Typically, the friction force and the electrostatic force are directed toward the target, while the electron temperature gradient and ion temperature gradient forces are directed away from the target. The friction force and the ion thermal force are dominant, and the parallel impurity velocity in a steady state is then given by

$$V_{Z\parallel}^{\text{imp}} = V_{i\parallel} + \frac{2.6Z^2\tau_s}{m_z}\frac{\partial T_\text{i}}{\partial s}. \tag{2.17}$$

Figure 2.11 shows a schematic view of a magnetic configuration in a poloidal single-null divertor tokamak, indicating the electrostatic force, friction force, and thermal force. Since the background plasma flow, V_i, is usually directed toward the lower divertor plates, the flow pushes the impurity downstream through friction. On the other hand, the parallel temperature gradient is directed upstream. In the open-flux surface, transport parallel to the magnetic field is more dominant than the transport perpendicular to the magnetic field. This is in contrast to the impurity profile in the closed magnetic field inside the SOL, which is determined by the transport perpendicular to the magnetic field. It is interesting to note that the temperature gradient has an opposite effect on impurity transport inside and outside the last closed-flux surface (LCFS). Although the temperature gradient

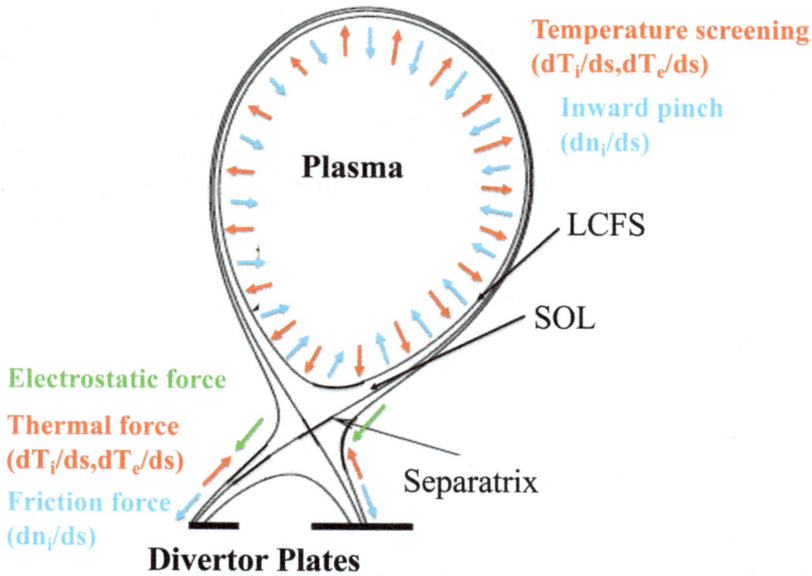

Figure 2.11. Schematic view of a magnetic configuration in a poloidal single-null divertor tokamak, with the electrostatic force, friction force, and thermal force indicated.

prevents the accumulation of impurities inside the LCFS via the temperature screening effect, it increases the impurity influx in the SOL and enhances the impurity concentration in the plasma core. Therefore, a larger temperature gradient perpendicular to the magnetic flux surface inside the LCFS and a smaller temperature gradient along the magnetic field in the SOL are preferable in order to reduce the concentration of impurities in a plasma.

Figure 2.12 shows schematics of the formation of a parallel flow toward divertor plates in an axisymmetric and a 3D divertor configuration. Here, the sign of parallel flow and toroidal angle ϕ represent the positive or negative toroidal angle. In the two-dimensional (2D) axisymmetric divertor configuration, the flux tubes are connected to the inner and outer divertor plates. Although the direction of parallel flow with respect to the toroidal angle is opposite between the inner and the outer divertor plates, these directions of parallel flow with respect to the plates are toward the divertor plates. In the 3D divertor configuration, the two flux tubes are connected to the same divertor plate and the directions of parallel flow with respect to the toroidal angle are opposite. However, the directions of parallel flow with respect to the plates are toward the divertor plates. The spatial separation between the counter-streaming flows (positive and negative parallel flows) are quite long in the 2D axisymmetric divertor configuration, and the pressure along the flux tube is generally conserved. In contrast, the spatial separation in the 3D divertor configuration is much shorter, and the parallel momentum loss of parallel flow via perpendicular viscosity occurs. In the 2D and 3D divertor configurations, the parallel flows are always toward the divertor plate, which contributes to impurity shielding at the boundary of the plasma.

In the 3D divertor configuration, there are many variations of the 3D magnetic configuration depending on the choice of optimization, as visualized in figure 2.13. These magnetic configurations can be categorized into Type I, a limiter configuration with a magnetic island; Type II, an island divertor; and Type III, an X-point poloidal (or helical) divertor with magnetic islands. The magnetic field toward the divertor plate has a highly complex structure, which becomes significant when the perturbation field is applied.

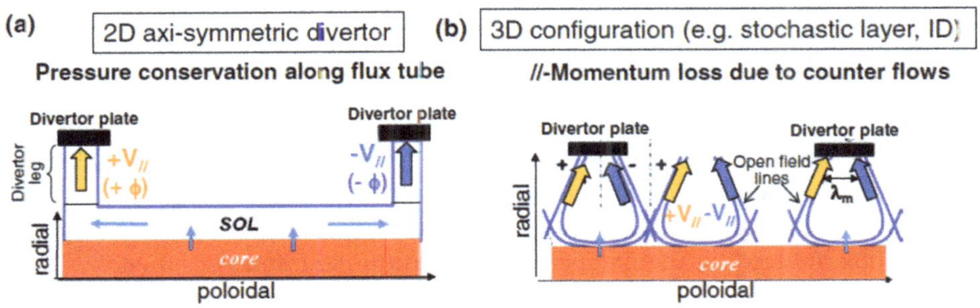

Figure 2.12. Schematics of the formation of a parallel flow toward divertor plates in (a) 2D axisymmetric, and (b) 3D divertor configuration. Reproduced from [22].

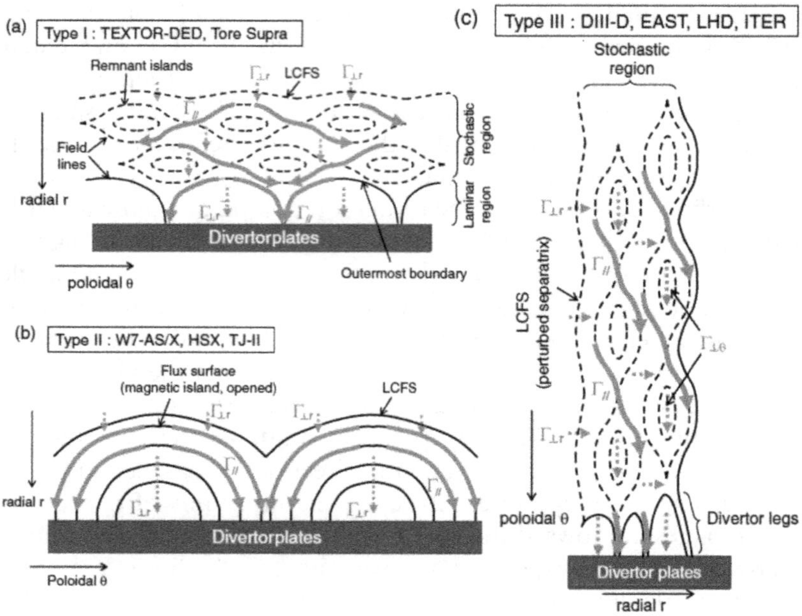

Figure 2.13. Schematics of divertor configurations and transport directions for (a) a limiter configuration with magnetic perturbation field (Type I), (b) an island divertor (Type II), and (c) an X-point poloidal/helical divertor with magnetic perturbation (Type III), respectively. Reproduced from [22].

- In Type I configurations, the perturbation field is applied in a limiter setup, and the transport between the LCFS and the divertor plates is dominated by the transport parallel to the stochastic field lines. The transport is considered to be affected by the stochastic layer due to radial projection of the field lines caused by magnetic field braiding. The magnetic configuration of Tokamak Experiment for Technology Oriented (TEXTOR)-DED and tungsten (W) Environment in Steady-state Tokamak (WEST, formerly Tore Supra) belong to this category.
- In Type II configurations, the magnetic field line of the islands across the divertor plate and the transport between the LCFS and divertor is dominated by the transport both parallel and perpendicular to the magnetic field line of the magnetic island. The magnetic configuration of W7-AS, W7-X, the Helically Symmetric Experiment (HSX), and TJ-II correspond to this type.
- In Type III configurations, the magnetic perturbation field is applied to a poloidal divertor configuration or helical divertor configuration, and a stochastic magnetic field layer is produced by the overlap of the magnetic islands induced by the magnetic perturbation field and the magnetic field by the helical coils. The transport is dominated by the poloidal component of the transport perpendicular to the field lines. The magnetic configuration of DIII-D, the Experimental Advanced Superconducting Tokamak (EAST), LHD, and ITER correspond to this type.

Transport in the open-flux surface region in a configuration with magnetic field perturbation is much more complicated than in the simple SOL without a perturbation field. In that case, the impurity screening effects strongly depend on the 3D structure of the magnetic field crossing the divertor plate.

References

[1] Liang Y, Ida K, Rice J E, Minami T, Funaba H and Kado S *et al* 2002 Observation of low impurity diffusivity inside the neoclassical transport barrier in the Compact Helical System *Phys. Plasmas* **9** 4179–87

[2] Connor J W 1973 The neo-classical transport theory of a plasma with multiple ion species *Plasma Phys.* **15** 765–82

[3] Hirshman S P and Sigmar D J 1981 Neoclassical transport of impurities in tokamak plasmas *Nucl. Fusion* **21** 1079–201

[4] Kovrizhnykh L M 1984 Neoclassical theory of transport processes in toroidal magnetic confinement systems, with emphasis on non-axisymmetric configurations *Nucl. Fusion* **24** 851–936

[5] Taylor J B 1961 Diffusion of plasma ions across a magnetic field *Phys. Fluids* **4** 1142–5

[6] Homma Y, Yamoto S, Sawada Y, Inoue H and Hatayama A 2016 Kinetic modelling for neoclassical transport of high-Zimpurity particles using a binary collision method *Nucl. Fusion* **56** 036009

[7] Wade M R, Houlberg W A and Baylor L R 2000 Experimental confirmation of impurity convection driven by the ion-temperature gradient in toroidal plasmas *Phys. Rev. Lett.* **84** 282–5

[8] Helander P, Newton S L, Mollén A and Smith H M 2017 Impurity transport in a mixed-collisionality stellarator plasma *Phys. Rev. Lett.* **118** 155002

[9] Perek A, Ida K, Yoshinuma M, Goto M, Nakamura Y and Emoto M *et al* 2017 Observation of the bulk ion density peaking in a discharge with an impurity hole in the LHD *Nucl. Fusion* **57** 076040

[10] Sudo S, Tamura N, Muto S, Funaba H, Suzuki C and Murakami A *et al* 2013 Transport characteristics of tracer and intrinsic impurities depending on the density of LHD plasmas *Plasma Phys. Control. Fusion* **55** 095014

[11] García-Regaña J M, Kleiber R, Beidler C D, Turkin Y, Maaßberg H and Helander P 2013 On neoclassical impurity transport in stellarator geometry *Plasma Phys. Control. Fusion* **55** 074008

[12] Pedrosa M A, Alonso J A, García-Regaña J M, Hidalgo C, Velasco J L and Calvo I *et al* 2015 Electrostatic potential variations along flux surfaces in stellarators *Nucl. Fusion* **55** 052001

[13] García-Regaña J M, Beidler C D, Kleiber R, Helander P, Mollén A and Alonso J A *et al* 2017 Electrostatic potential variation on the flux surface and its impact on impurity transport *Nucl. Fusion* **57** 056004

[14] Mollén A, Landreman M, Smith H M, García-Regaña J M and Nunami M 2018 Flux-surface variations of the electrostatic potential in stellarators: impact on the radial electric field and neoclassical impurity transport *Plasma Phys. Control. Fusion* **60** 084001

[15] Fülöp T and Nordman H 2009 Turbulent and neoclassical impurity transport in tokamak plasmas *Phys. Plasmas* **16** 032306

[16] Fülöp T, Braun S and Pusztai I 2010 Impurity transport driven by ion temperature gradient turbulence in tokamak plasmas *Phys. Plasmas* **17** 062501

[17] Skyman A, Nordman H and Strand P 2012 Impurity transport in temperature gradient driven turbulence *Phys. Plasmas* **19** 032313

[18] Ida K and Fujita T 2018 Internal transport barrier in tokamak and helical plasmas *Plasma Phys. Control. Fusion* **60** 033001

[19] Mikkelsen D R, Tanaka K, Nunami M, Watanabe T H, Sugama H and Yoshinuma M *et al* 2014 Quasilinear carbon transport in an impurity hole plasma in LHD *Phys. Plasmas* **21** 082302

[20] Stangeby P C 2000 *Boundary of Magnetic Fusion Devices* (Boca Raton, FL: CRC Press) 1st edn

[21] Braginskii S I 1965 Transport processes in a plasma *Rev. Plasma Phys.* **1** 205 (transl. from Russian, originally published 1963)

[22] Kobayashi M, Xu Y, Ida K, Corre Y, Feng Y and Schmitz O *et al* 2015 3D effects of edge magnetic field configuration on divertor/scrape-off layer transport and optimization possibilities for a future reactor *Nucl. Fusion* **55** 104021

Katsumi Ida and Naoki Tamura

Chapter 3

Diagnostics

In this chapter, diagnostic instruments utilized in magnetically confined fusion (MCF) experiments, in particular for the study of impurity transport, are described. As already described in the previous chapter, the control of impurities in MCF plasmas is a high-priority issue in the development of magnetic fusion reactors. To this end, it is extremely important to know firstly the composition of impurities in MCF plasmas. This is critical to avoid fuel dilution by the impurities. Secondly, it is very important to know the spatio-temporal behavior of the impurities in MCF plasmas. The transport properties of the impurities, which can be derived from such spatio-temporal behavior, and the underlying physical mechanisms are essential information in order to control the impurities.

3.1 Introduction

Impurity ions and atoms, as well as hydrogen isotope ions, in MCF plasmas emit various line radiation together with continuum radiation. The wavelengths of such radiations are spread over a wide spectral range, across the near-infrared (NIR), visible (VIS), ultraviolet (UV), vacuum ultraviolet (VUV), extreme ultraviolet (EUV), and soft X-ray (SXR) domains. The definitions of the spectral regions discussed in this book are summarized in table 3.1 [1].

Both the line and continuum radiations from impurity ions and atoms contain various information on the impurities present in MCF plasmas. The line radiation arises from a transition, e.g., from an energy level j to energy level k, of an atom or ion (sometimes molecules or atomic nuclei, also). In high-temperature MCF plasmas, electron-impact excitation and de-excitation and radiative decay are dominant in the energy-level-population mechanisms of the atoms or ions. In reality, other level-population mechanisms, such as ion-impact excitation and de-excitation and charge-exchange recombination, should be taken into account according to the situation. Of these, charge-exchange recombination is the basis for active spectroscopy, which will be discussed in detail in section 3.3. Several

doi:10.1088/978-0-7503-1451-0ch3

Table 3.1. Spectral regions relevant to spectroscopy of magnetically confined plasmas. [1] Taylor & Francis Ltd. http://tandfonline.com.

Spectral Region	Wavelength/Energy Range
Near-infrared (NIR)	700 to 1200 nm/1 to 2 eV
Visible	400 to 700 nm/2 to 3 eV
Ultraviolet (UV)	200 to 400 nm/3 to 6 eV
Vacuum ultraviolet (VUV)	30 to 200 nm/6 to 40 eV
Extreme ultraviolet (EUV)	10 to 30 nm/40 to 120 eV
Soft X-ray (SXR)	0.1 to 10 nm/120 to 12 000 eV

models have been proposed and applied to understand line radiation in plasmas. For instance, a corona model has been developed for cases in which the line radiation is emitted from a plasma with a low density and high temperature. Meanwhile, for cases in which the line radiation is emitted from a plasma with a high density, a local thermal equilibrium (LTE) model has been developed. In reality, neither model can explain correctly line radiation from plasmas, especially MCF plasmas. As for the LTE model, near-valid conditions have been confirmed in a cold plasma, the so-called 'ablation cloud', which surrounds a pellet that is injected into the plasma from outside for the purpose of fueling and diagnostics. In order to fully comprehend the line radiation from actual MCF plasmas, the possible energy-level transitions, such as the impact excitations and de-excitations and recombinations of atoms or ions, should be taken into account, which is the basis of the collisional-radiative (CR) model. By using the CR model, the density of ions in each charge state of the impurity can be estimated. However, because it is impossible to consider all the possible transitions of all the possible levels, reasonable assumptions on the transitions and levels included are made depending on the plasma conditions.

The spectral line shape has a finite width due to several atomic processes and transport. In plasmas with a low density and a high temperature, the width of the spectral line is usually determined by Doppler broadening. In cases with plasmas with a high density, such as an ablation cloud, the width of the spectral line is determined by another mechanism, Stark broadening. The full width half maximum (FWHM) of a spectral line with Doppler broadening is given by

$$\lambda_{\mathrm{FWHM}} = 7.68 \times 10^{-5} \lambda_0 \sqrt{\frac{T_{\mathrm{i}}}{A}}, \tag{3.1}$$

where λ_0, T_{i}, and A are the original central wavelength of the line emission, the ion temperature (in eV), and the atomic mass number of the emitting species, respectively. By using the shift from the central wavelength of the spectral line, the flow velocity of the emitting species can be obtained.

In continuum radiation, liberated (free) electrons play an important role. Transitions between two continuum states of electrons result in bremsstrahlung, and transitions between the continuum state and the bound state of electrons result in recombination radiation. The total bremsstrahlung power density per unit energy interval (W m^{-3} eV) can be expressed as

$$\frac{dP_{\rm ff}^{\rm tot}}{dE} = 1.54 \times 10^{-38} n_{\rm e}^{2} Z_{\rm eff} \bar{g}_{\rm ff} \frac{e^{-E/T_{\rm e}}}{\sqrt{T_{\rm e}}}, \tag{3.2}$$

where $n_{\rm e}$ is the electron density, $T_{\rm e}$ is the electron temperature, E is the photon energy, $Z_{\rm eff}$ is the effective ion charge, and $\bar{g}_{\rm ff}$ is the free–free transition Gaunt factor averaged over a Maxwellian electron velocity distribution. The effective ion charge is formulated as

$$Z_{\rm eff} = \sum_{Z} \frac{n_Z Z^2}{n_{\rm e}}, \tag{3.3}$$

where Z is the ion charge. The factor of $Z_{\rm eff}$ can be used as a measure of the impurity concentration (i.e., the increment from the baseline, where $Z_{\rm eff}$ is a unity for the pure hydrogen plasmas). The factor of $\bar{g}_{\rm ff}$ can be studied in different spectral regions (e.g., $\bar{g}_{\rm ff} \sim 1$ for the SXR region, $\bar{g}_{\rm ff} \sim 2$–5 for the visible region). The recombination power density per unit energy can be expressed in almost the same manner, but more complex evaluations on possible transitions are necessary. According to the law of the conservation of energy, in the recombination process, when the energy of the free electrons is less than the ionization energy from the level where the free electrons are captured, light is not emitted. This means that the recombination radiation is negligible in the visible domain (in contrast, recombination radiation in the SXR domain dominates the continuum radiation). Then, the continuum light in the visible region is purely attributed to bremsstrahlung. Thus, a measurement of continuum light intensity in the visible domain can provide information regarding $Z_{\rm eff}$, i.e., the concentration of impurities in the plasma.

As mentioned above, spectroscopic analysis can provide various information on the impurities in MCF plasmas. The different methods of spectroscopic measurement can be categorized as either passive or active. In passive spectroscopy, natural emissions from the impurity ions or atoms in the plasma are usually measured using spectroscopic instruments. By contrast, in the active spectroscopy, external operations are performed to produce or enhance emissions from the impurity ions or atoms. Thus, active spectroscopy grants the possibility of measuring more accurate (e.g., local quantities) information. In the following sections, both passive and active spectroscopy for the study of impurity transport will be introduced.

3.2 Passive spectroscopy

3.2.1 UV–visible–NIR spectroscopy

The plasma density and temperature determine the fractional abundance of the impurity ions and the wavelengths of their light emissions. From the edge and divertor region of magnetically confined toroidal plasmas, emissions in the UV, visible, and NIR domains are usually dominant. Therefore, measurements of UV, visible, and NIR emissions in MCF experiments are generally utilized for (i) the identification and quantification of impurities from plasma-facing components (PFCs), (ii) the characterization of the spatio-temporal behavior of the impurities in such edge and divertor region. A significant advantage of measurement of UV, visible, and NIR emissions is that the measurement can be performed in atmospheric environments. A UV-grade fused silica window has a more than 90% transmittance for the emissions in such a spectral region (i.e., between 200 nm and 1200 nm). In such setups, the measuring instruments (e.g., photodetectors, cameras, and spectrometers) can be arranged and exchanged in a flexible manner outside the vacuum vessel by attaching such windows as an interface to the vacuum vessel. Another advantage of measurement in the UV, visible, and NIR domains, especially in the visible and NIR, is that an optical fiber can be used to transmit the light collected by the optics located near the optical window mounted on the vacuum vessel to the instruments for measurement. A typical attenuation (in other words, transmission loss) of a high-purity fused silica optical fiber is around 30 dB km^{-1} at a wavelength of 400 nm. Attenuation below a wavelength of 300 nm increases significantly. In such cases, in large MCF experimental devices such as the Joint European Torus (JET) and Large Helical Device (LHD), measurement of the UV emission (200 to 300 nm) should be performed in the vicinity of the devices, because the necessary length of the optical fibers between the windows and the measurement instruments in the refuge room is typically around 100 m. The numerical aperture of the optical fibers used in the MCF experiments is typically around 0.22, which corresponds to an F number of 2.2. Polymer-clad silica fibers, which consist of an optical core made from fused silica and an optical cladding made of polymer, have also been utilized in MCF experiments. Recently, a plastic optical fiber (POF), which is made completely from polymer, has been developed for the purpose of telecommunications. Its robustness is one of its main advantages over silica-core optical fibers. However, it should be noted that the attenuation of the POF is around 1 dB m^{-1} at a wavelength of 650 nm. The utilization of optical fibers bring another benefit to the diagnostics layout of MCF experiments, because the immediate vicinity of MCF experimental devices is a harsh environment with high magnetic field and the emittance of neutrons and gamma rays.

3.2.1.1 Filtered detectors and filtered cameras

In order to detect simply the discrete visible line emissions from the edge and divertor region, a filtered photodetector-based measurement system has been widely utilized in many MCF experimental devices, such as DIII-D, the Korea Superconducting Tokamak Advanced Research (KSTAR), the Experimental Advanced Superconducting Tokamak (EAST), the Wendelstein 7-X (W7-X), and others. This system is commonly referred to as a 'filterscope' [2]. This fact attests that filterscopes or functionally equivalent systems have been installed on almost all MCF experimental devices. Usually, a photomultiplier tube (PMT) is used as a photodetector for the filterscope, due to its high sensitivity.

Figure 3.1 shows a schematic diagram of the filterscope system on the EAST tokamak. The light collected by an optical lens from the plasma is focused onto the entrance part of an optical fiber bundle. The light is then transferred to a

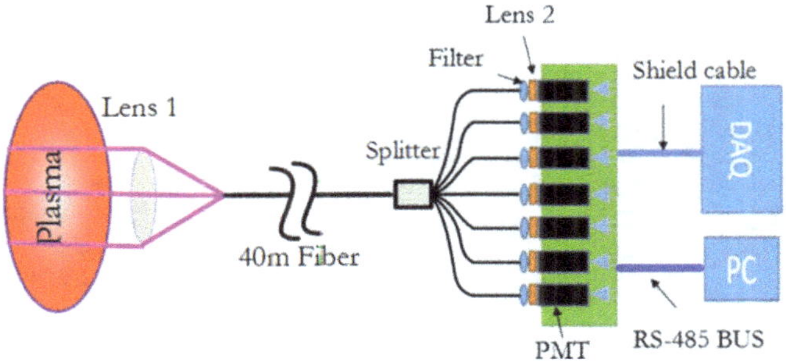

Figure 3.1. Schematic diagram of the filterscope system on the EAST tokamak. Reprinted from [3], with the permission of AIP Publishing.

low-radiation and low-field diagnostic area by each optical fiber. The length of the fiber bundle is 40 m. In the diagnostic area, the light conveyed by each optical fiber is divided by an optical splitter into seven PMT modules. This configuration enables one to measure the different visible line emissions using the same sight line. The core diameter and numerical aperture of each optical fiber used in the EAST filterscope system is 600 μm and 0.22, respectively. The spatial resolution obtained is about 2 cm. Before entering the PMT modules, the light is filtered by an optical bandpass filter and is collimated by a lens. In the EAST tokamak, D Balmer-α (656.1 nm), D Balmer-γ (433.9 nm), He II (468.5 nm), Li I (670.8 nm), Li II (548.3 nm), C III (465.0 nm), O II (441.5 nm), Mo I (386.4 nm), W I (400.9 nm), and visible bremsstrahlung radiation (538.0 nm) are monitored using the filterscope system. The bandpass filter for each wavelength has a FWHM of 5 nm and a transmittance of over 70%. In total, 125 channels are available in the EAST filterscope system. All the signals from the PMT modules are digitized up to 200 kHz simultaneously. The high-time-resolved optical signals can be also used for the study of edge magneto-hydrodynamic (MHD) instabilities, such as the edge-localized mode.

The Tunsten(W) Enviroment in Steady-state Tokamak (WEST) also utilizes a filterscope, developed by the Oak Ridge National Laboratory (USA), to measure the line emissions of the impurities that originate from the PFCs. The left panel of figure 3.2 shows the lines of sight (16 channels in total) of the filterscope looking at the outboard limiter from the inboard side in the WEST vacuum vessel. The spatial resolution of the filterscope for the outboard limiter is 90 mm. As shown in figure 3.2, spatial profiles of the W I and Mo I emissions are successfully obtained with the filterscope. The W I profile is found to be less peaked compared to the Mo I profile. This experimental result suggests that the runaway electrons induced more damage (i.e., increased the source of impurities) in the equatorial plane location.

A filtered visible camera was used to monitor plasma conditions as a snapshot in early MCF experiments. Since the advent of fast framing cameras, a filtered camera diagnostic has been widely utilized to track fast transitional phenomena, plasma filamentary behaviors, etc., in many MCF experimental devices, such as the Alcator C-MOD, the Compact Assembly (COMPASS), DIII-D, the National Spherical Torus Experiment - Upgrade (NSTX-U), the LHD, and W7-X. Fast visible

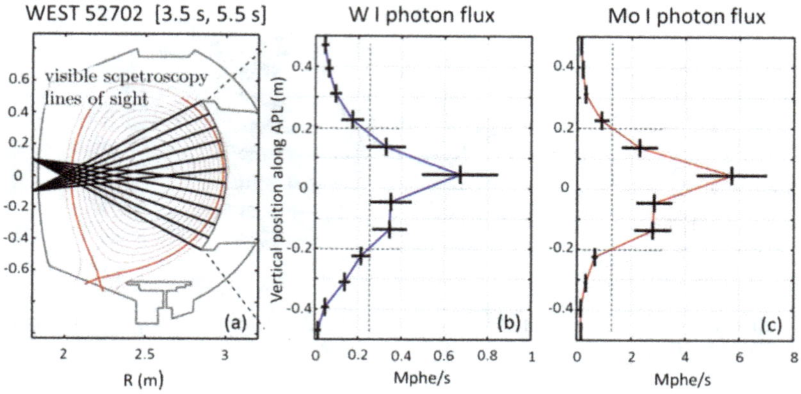

Figure 3.2. Filterscope lines of sight for the outboard limiter in the WEST tokamak (left), and the W I (middle, blue) and Mo I (right, red) radiances along the outboard limiter. Reprinted from [4], with the permission of AIP Publishing.

diagnostics are also used to monitor detailed information of the impurities injected artificially from outside. Such information (e.g., time and depth of the penetration of impurities injected) is highly important in performing transport modeling of impurity injection using a simulation code as inputs. In some cases, different fields of view (e.g., looking from above/behind and tangentially) are utilized to obtain such information.

As a common feature, the central wavelength of the dielectric interference filter will shift to a lower wavelength with an increase in the incident angle. The degree of the wavelength shift is dependent upon the incident angle and the effective refractive index of the filter, which can be written by the following formula [5]:

$$\lambda_\alpha = \lambda_0\sqrt{(1 - \left(\frac{N_e}{N*}\right)^2 \sin^2\alpha)}, \tag{3.4}$$

where λ_α is the center wavelength at the angle of incidence, λ_0 is the center wavelength at normal incidence, N_e is the refractive index of the external medium (air = 1), $N*$ is the effective refractive index of the filter, and α is the angle of incident light. This feature is very useful in tuning a very narrow bandpass filter to the desired central wavelength. However, in the case of filtered camera diagnostics, it requires cautious use. Bandpass filters for imaging applications require a broad bandpass feature, and they should be placed where the rays of light are almost parallel.

3.2.1.2 Visible grating spectrometers

Monochromators/spectrometers have been extensively utilized in order to measure effectively the spectral components of light from plasma. A monochromator can output a selectable narrow band of wavelengths of light, while a spectrometer can output a selectable broad band (i.e., spectrum) of wavelengths of light. Here, we take the spectrometer as an example. A spectrometer uses a dispersing element to resolve light into its spectral components. There are two basic types of dispersing elements: a prism or diffraction grating (a so-called grism is the combination of a prism and diffraction grating so that light at a central selected wavelength passes straight through it). According to their shape, diffraction gratings can be classified into plane, spherical-concave, or toroidal-concave types. Further, according to the direction of light scattered by the grating, diffraction gratings can be classified as being either reflective or transmissive. In general, spectrometers with a reflective grating are more widely used than those with a transmission grating. A dispersion scheme using the diffraction grating is also utilized for the measurement of light in the VUV, EUV, and SXR regions. The grating substrate is ruled with a large number of grooves, which can be either equally spaced or variably spaced, to reduce aberrations in a concave grating [6] and for self-focusing in a plane grating [7].

In order to observe the behavior of impurities in a plasma with a good time resolution, which should be faster than the typical impurity confinement time, the output spectrum from a spectrometer should be measured with a high spectral and temporal resolution. To this end, a multi-channel detector, e.g., a photodiode array (PDA) or charge-coupled device (CCD), is widely used in the spectrometer for MCF experiments. The typical channel numbers of such multi-channel detectors are currently from 512 to 2048 channels, and the typical temporal resolution is 10 ms or less. When a millisecond order or less is needed as the temporal resolution of measurement using the spectrometer, an additional piece of equipment is required, because the intensity of light exposed to the PDA or the CCD for such a short time will be very weak. In such a case, an image intensifier (I.I.) is often additionally installed to increase the intensity of the available light in the spectrometer. An image intensifier is a vacuum-tube device, which typically consists of a photocathode, a multi-channel plate (MCP), and a phosphor screen. The photocathode has a function converting photons into electrons. By applying a strong electric field (several hundreds of volts) across the MCP, the MCP can multiply the electrons generated by the photocathode. Finally, the multiplied electrons are converted back into photons by the phosphor screen. The multiplication factor can be increased by increasing the number of MCP stages (normally up to three stages). One of the disadvantages of using an I.I. is that the width of the spectral lines is broadened due to the multiplication process in the MCP. Thanks to the development of semiconductor fabrication technology, the development of technology regarding solid-state image-sensing devices is proceeding apace, first with the development of front-side illuminated (FI) CCD chips, and later with back-side illuminated (BI) CCD chips. Owing to their structural advantages, BI-CCD chips have a much higher quantum efficiency (80% or more as a peak value in the sensitive wavelength region); in other words, a much higher sensitivity than traditional FI-CCD chips. Consequently, in some cases, a BI-CCD detector can take the place of a combined I.I. and multi-channel detector. With this, high-sensitivity spectroscopic systems with good spectral resolution become possible. Recently, an electron-multiplying CCD (EMCCD) chip has been developed for faint-light detection with a high signal-to-noise ratio. In the EMCCD chip, there are two readout registers, a conventional register and an electron-multiplying (EM) register. Through the EM register, a significant gain factor (up to around 1000 times) can be obtained. Since this amplification process occurs before the signal readout, causing the readout noise, a high signal-to-noise ratio can be achieved. Moreover, significant developments in semiconductor technology have also improved the performance of traditional complementary metal-oxide semiconductor (CMOS) chips. Recent scientific CMOS (sCMOS) chips have a comparable (in some cases, unrivaled) ability against up-to-date CCD chips. Therefore, it is necessary to choose the most suitable detector according to the situation. When a much higher time resolution, such as of microsecond order or less, is required, a single-channel detector, such as a PMT or a photodiode, is typically used to measure a single spectral line emission.

The basic grating equation can be written as

$$m\lambda = d(\sin\alpha + \sin\beta), \tag{3.5}$$

where λ, d, m, α, and β are the wavelength of the light, the groove spacing, the spectral order number, and the angles of incidence and diffraction measured with respect to the normal of the grating surface, respectively. The groove density of the grating, $1/d$, is usually defined as the number of grooves per millimeter. Although there are many different possible optical configurations for a grating spectrometer, the following three optical configurations are the most common: crossed Czerny–Turner, unfolded Czerny–Turner, and concave holographic spectrometers. Figure 3.3 shows an example of a spectrometer in a crossed Czerny–Turner configuration, which is used for the measurement of Z_{eff}, an important measure of the impurity concentration, in the LHD [8]. In this configuration, a toroidal mirror and a flat mirror collimates light from an entrance slit onto the grating, and two spherical mirrors focus the diffracted light from the grating onto the image plane of the CCD detector. A short focal length of 300 mm is adopted to achieve an extremely bright system. Another example of a Czerny–Turner–like spectrometer [9] for charge-exchange recombination spectroscopy (CXRS) is shown in figure 3.4. The details of this CXRS will be explained in section 3.3. In this configuration, in order to parallelize the light onto the grating (128 × 154 mm with 2160 grooves mm^{-1}), a camera lens with an F number of 2.8 and a focal length of 400 mm is used at the entrance to the spectrometer instead of mirrors. At the exit, in order to refocus the diffracted light onto the image plane of the CCD detector, the same camera lens is used. Utilizing a camera lenses with a small F number results in a high-throughput visible spectrometer, which is needed for high-time-resolved charge-exchange spectroscopy (or CXS). As can be seen in figure 3.4, a special fiber bundle, in which the fibers are arranged in two columns, is coupled with the entrance camera lens. This is done to increase the number of channels captured on a single CCD. In order to prevent overlapping of the spectra from the two columns of fibers, an interference filter is installed on the camera lens for diffracted light. To achieve a high-throughput spectrometer for high-time-resolved CXRS, spectrometers with transmission gratings have also been utilized. The grating used in these cases is called a volume phase holographic (VPH) transmission grating. Recently, by using a VPH transmission grating sandwiched between two prisms (i.e., a grism), transmission grating spectrometers with a higher dispersion have been realized.

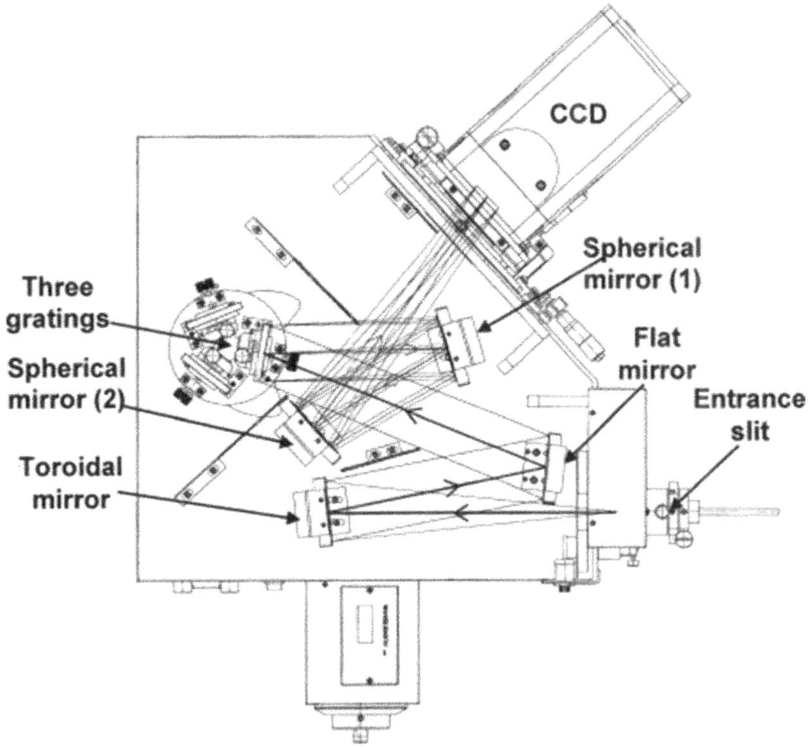

Figure 3.3. Optical arrangement of 300 mm crossed Czerny–Turner-type visible spectrometer. The light path from the entrance slit is indicated by a black arrow. Reprinted from [8], with permission from The Japan Society of Plasma Science and Nuclear Fusion Research.

As already shown in figures 3.3 and 3.4, utilizing toroidal and spherical mirrors or lenses in a Czerny–Turner configuration provides a sufficient degree of correction to the spherical aberration and astigmatism. Consequently, such astigmatism-corrected spectrometers can provide a good spectral image for the CCD detector for each fiber of the fiber array aligned vertically at the entrance slit of the spectrometer. Figure 3.5 shows an example of a two-dimensional (2D) measurement of impurity flow velocity based on Doppler shift analysis for the spectra obtained with a space-resolved visible spectroscopic system in the scrape-off layer (SOL) region of the LHD [10]. This space-resolved visible spectroscopic system has a total of 130 channels to obtain a 2D distribution of the emission spectra in the SOL region of the LHD. The spectrometer employed here has a rotatable table to facilitate the exchange of gratings with different groove densities (300, 1800, and 2400 grooves mm^{-1}). By applying the 2400 grooves mm^{-1} grating, the Doppler shift of the line emission from the impurity ions can be evaluated, making it possible to evaluate the flow velocity of the impurity ions in the SOL region of the LHD. As can be seen in figure 3.5, the C^{2+} ions on the left-hand side of the X-point went in a positive direction, which is consistent with the fact that the divertor legs on the left-hand side of the X-point are

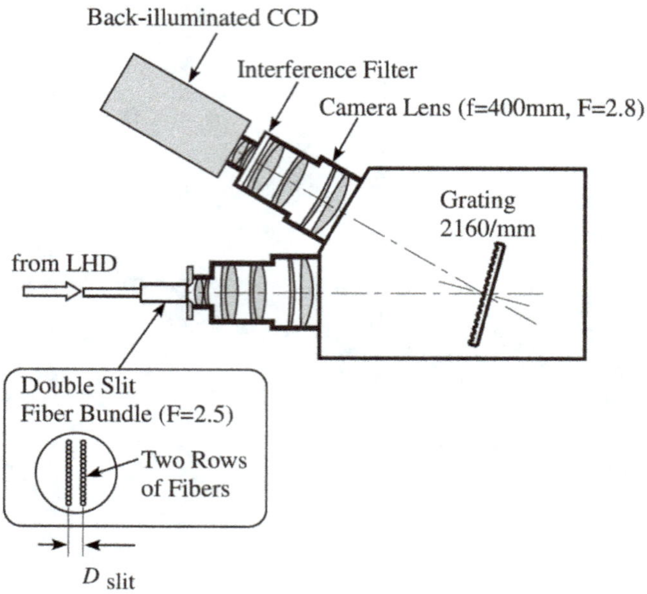

Figure 3.4. Top-down view of a Czerny–Turner–like spectrometer using camera lenses for charge-exchange recombination spectroscopy. [9] Taylor & Francis Ltd. http://tandfonline.com.

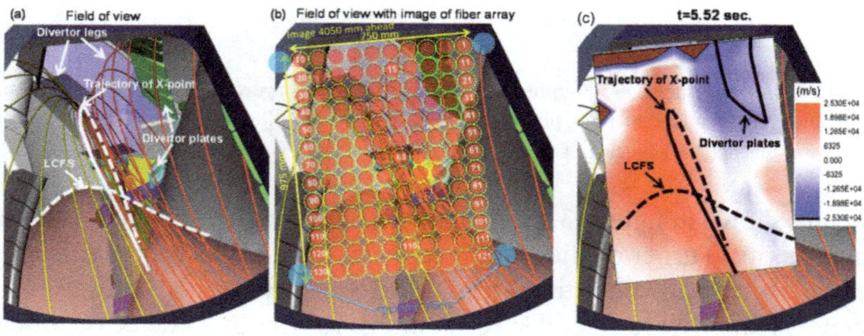

Figure 3.5. Example of a two-dimensional measurement of impurity flow velocity based on Doppler shift analysis for spectra obtained with a space-resolved visible spectroscopic system in the SOL region of the LHD. (a) Field of view (FOV) of the fiber array with magnetic field structure. The red and yellow lines represent magnetic field lines of the divertor legs connected to the bottom and upper divertor plates. (b) Images of the 130-channel fiber array are overlaid in the FOV. (c) Two-dimensional distribution of flow velocity of C III (C^{2+}) estimated from the Doppler shift analysis, at t = 5.52 s of LHD shot #128200. Red and blue colors represent the positive and negative directions of the flow. Reprinted from [10], with the permission of AIP Publishing.

connected to the divertor plate located at the front side of the paper. Meanwhile, the divertor legs on the right-hand side of the X-point are connected to the divertor plate located at the back side of the paper. Consequently, a negative flow velocity of the C^{2+} ions on the right-hand side is also reasonable.

3.2.2 VUV and EUV spectroscopy

When the electron temperature in the core region of an MCF plasma rises above 1 keV, light emissions from the impurity ions are present in the VUV, EUV, and SXR domains. Therefore, in MCF experiments, spectroscopic analyses in the VUV/EUV/SXR regions are most commonly performed to study the behavior of impurities in the core MCF plasma. We begin by detailing the available techniques for the measurement of light in the VUV/EUV domains.

3.2.2.1 Principle of VUV and EUV spectroscopy

Fluoride glasses, such as magnesium fluoride (MgF_2) and lithium fluoride (LiF), are widely utilized as window materials for UV light. Due to their large bandgaps (about 10 eV), fluoride glasses can also transmit VUV light, the wavelength of which is above around 100 nm (still not far from the UV domain). For light with a wavelength of below 100 nm, there is no transmissive elements because such VUV/EUV light is mostly absorbed in matter. Therefore, transmissive (refraction) elements, like interference filters for visible light, are almost infeasible for light with a wavelength between 10 nm and 100 nm. As described in the previous section, dispersion techniques with a diffraction grating can also be utilized for the measurement of VUV/EUV emissions. However, the reflectivity of metals and dielectric coatings for VUV/EUV emissions drops significantly, which leads to very low efficiency for the spectrometers requiring many reflections. Accordingly, the spectrometers measuring VUV/EUV emissions often use a concave grating, which can act as both a diffracting and focusing element, i.e., can reduce the number of reflections. For light with a wavelength above about 50 nm, concave grating spectrometers can operate at near-normal incidence because the reflectivity of the coated gratings are typically >50%. On the other hand, for light with shorter wavelengths, since the normal-incidence reflectivity of the coated gratings is very low, spectrometers are designed to operate at grazing-incidence angles of about 85°. In each case, a vacuum condition inside the spectrometer is essential, because the transmittance of light for which the wavelength is less than 200 nm falls sharply in air, mainly due to the absorption by molecules of oxygen and nitrogen.

When one designs an optical layout for a spectrometer using a concave grating, a Rowland circle concept is often used. In the concept of a Rowland circle, if the concave grating is placed tangentially on the circumference of a circle with a diameter R equal to the radius R of the curvature of the concave grating, then the point of the light source (i.e., the entrance slit) and the detection point (or the exit slit) should also be on the circumference of the same circle. Therefore, a simple concave grating spectrometer based on a Rowland circle always suffers from a significant astigmatism. To compensate for this astigmatism in concave grating spectrometers, an aberration-corrected toroidal-concave grating and a concave grating with variably spacing grooves have been developed.

The spectrally resolved VUV/EUV emissions obtained by grating spectrometers can be measured with almost the same schemes as the visible grating spectrometers. PDAs or CCDs are often set directly at the detection point (plane) of the grating

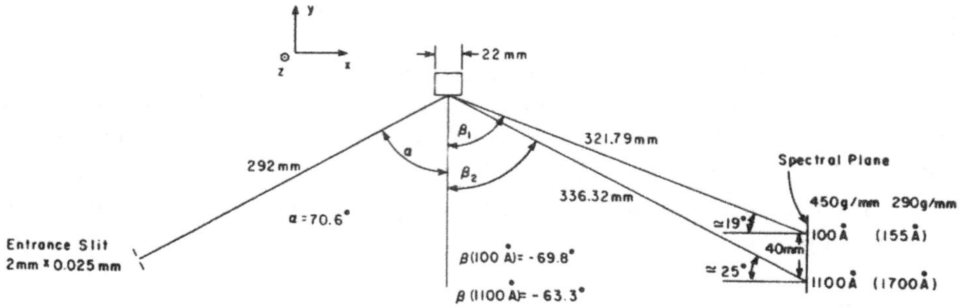

Figure 3.6. Optical schematic of a toroidal grating flat-field spectrometer, SPRED. Reprinted from [11]. © The Optical Society.

spectrometer, because they have a sensitivity to light in the VUV/EUV domains. In this case, several features of the PDAs or CCDs should be noted: (i) an input window can limit the observable wavelength region of PDAs and CCDs, (ii) coatings on the detectors' surface can also limit the sensitivity of the PDAs and CCDs, and (iii) the spectral sensitivity of the PDAs and CCDs is not flat in the VUV/EUV domain. From these points of view, a vacuum-compatible BI-CCD detector without any coatings is currently the most suitable detector design for VUV/EUV spectrometers. However, when one has to use a FI-CCD detector with an input window, one can utilize the conversion technique utilizing both an MCP and a phosphor screen. After the phosphor screen, such a CCD detector can be installed via an optical relay system using lenses, a fiber optic plate (FOP), or an imaging fiber bundle. This VUV/EUV-to-visible-light conversion technique with an MCP and phosphor screen is also utilized for high-time-resolved VUV/EUV measurement, because a higher gain is expected by utilizing the MCP. The MCP can detect not only single particles, such as electrons, ions and neutrons, but also photons in the UV and X-ray regions. However, the input surface of the MCP is usually coated with a photocathode, such as cesium iodide (CsI) or potassium bromide (KBr), to improve the quantum efficiency.

3.2.2.2 VUV and EUV grating spectrometers

An aberration-corrected toroidal grating spectrometer with a flat focal field design, which is referred to as a Survey, Poor REsoultion, extended Domain (SPRED) spectrometer [11], is widely used in MCF experiments for measurement of VUV/EUV emissions. Figure 3.6 shows an optical schematic of such a SPRED spectrometer. Light with an incident angle of 70.6° is directed toward the toroidal grating, which gives a good reflectivity of the light with a wavelength of down to 10 nm. Owing to its unique aberration-corrected toroidal grating feature, the astigmatism is almost zero even when a point source is set at the top of the entrance slit (i.e., 1 mm off from the optical axis). Consequently, a high-throughput VUV/EUV spectrometer is achieved. Figure 3.7 shows an example of the spectrum obtained by the SPRED spectrometer in experiment with the JET tokamak [12]. The SPRED spectrometer installed in the JET tokamak has a 450 grooves mm^{-1} grating,

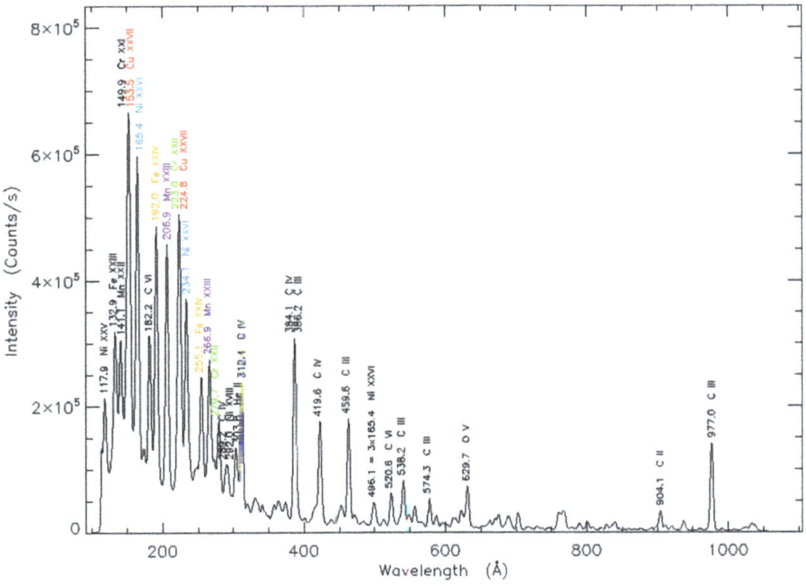

Figure 3.7. The SPRED spectrum for JET pulse #59385 averaged between times 6.1 and 7.0 s. Reproduced from [12].

resulting in a spectral range from about 100–1100 Å with a spectral resolution of about 5 Å. As can be seen in figure 3.7, intense line emissions from intrinsic impurities such as chromium (Cr), manganese (Mn), iron (Fe), nickel (Ni), and copper (Cu) are clearly observed. In the JET SPRED spectrometer, for high-spectral-resolution measurement a 2105 grooves mm^{-1} grating has additionally been installed, like the SPRED upgrade (which can also cover a spectral range of 100–320 Å with ~0.4 Å resolution) for the Tokamak Fusion Test Reactor (TFTR) and PBX tokamaks.

As one may have already grasped, there is a trade-off between observable spectral range and attained spectral resolution. Therefore, in order to cover the entire spectral range in the VUV/EUV domains of interest, more than one VUV/EUV spectrometer can be installed in MCF experimental devices. In the W7-X stellarator, four sets of flat-field spectrometers based on an aberration-corrected holographic toroidal grating are installed, which is known as the High-Efficiency eXtreme ultraviolet Overview Spectrometer (HEXOS) system. The HEXOS covers a wavelength range from 2.5 to 160 nm in total. One of its unique features is a high time resolution of 1 ms. The combination of a CsI-coated open MCP and an N-channel metal-oxide semiconductor (or NMOS) linear array detector enables the spectra obtained with the HEXOS system to be recorded with such a high time resolution.

We provide the following examples of VUV/EUV spectrometers with a varied-line-spacing (VLS) diffraction grating. The Lawrence Livermore National Laboratory electron-beam ion-trap facility has developed high-resolution VUV/EUV spectrometers also for MCF experiments. For the NSTX-U spherical tokamak, three VUV/EUV spectrometers, the X-ray Extreme Ultraviolet spectrometer (XEUS), the Long-Wavelength Extreme Ultraviolet Spectrometer (LoWEUS), and the Metal Monitor and Lithium Spectrometer Assembly (MonaLisa), have been installed. The wavelength coverage of the XEUS, LoWEUS, and MonaLisa is roughly 0.8–7 nm, 19–44 nm, and 5–22 nm, respectively. These wavelength ranges, of course, can be changed according to the intended use. The averaged line spacing of the VLS grating for the XEUS, LoWEUS, and MonaLisa are 2400, 1200, and 1200 grooves mm^{-1}, respectively. The XEUS and LoWEUS have also been installed on the Alcator C-MOD and DIII-D tokamaks. Chowdhuri and Morita *et al* have also developed two EUV spectrometers with a VLS grating, the EUV_Short and EUV_Long, for the LHD heliotron. The wavelength coverage of the EUV_Short and EUV_Long is 1–13 and 5–50 nm, respectively. These types of EUV spectrometers have also been installed on the Huan-Liuqi-2A (HL-2A) and EAST tokamaks.

In order to understand the spatio-temporal behavior of the impurities in MCF plasmas, spatial profiles of the VUV/EUV emissions from the impurity ions in MCF plasmas have also been constructed. In the earliest stage of such a measurement, a radial profile of the VUV emission is scanned by means of a rotating mirror installed in the vacuum vessel of the MCF experimental device (e.g., the ASDEX-Upgrade tokamak, the TJ-II heliac, and the CHS heliotron/torsatron). As 2D detectors, such as CCDs, have become more readily available, the radial profiles of VUV emission can be observed based on the imaging properties of normal- and near-grazing-incidence spectrometers with such detectors and a height-limited entrance slit. Such space-resolved VUV measurements have been performed in the LHD heliotron and HL-2A tokamak. Regarding space-resolved EUV measurement, as a simplest means, a radial profile of the EUV emission can be obtained by moving the spectrometer, connected to the vacuum vessel through bellows, up and down with a hydraulic jack. This method requires a shot-to-shot basis measurement or a long-duration discharge in order to obtain a spatial profile of the VUV/EUV emissions from the MCF plasma. In fact, this method requires a high reproducibility of the plasma, including with regards to the impurities. In order to solve these issues regarding reproducibility, Dong *et al* [13] developed a space-resolved EUV spectrometer by utilizing a height-limited entrance slit and a large-curvature grating. Figure 3.8 shows a schematic top view of the space-resolved EUV spectrometer and an optical schematic of the spectrometer in a vertical direction. A BI-CCD detector is used as the detector, the position of which can be moved along the focal plane by using an electric motor to change the observable spectral region. One of the key elements in the space-resolved EUV spectrometer is a spatial resolution slit, which is installed just in front of the entrance slit. The width of the entrance slit can be chosen between 30 and 100 μm, and the width of the spatial resolution slit can be set as either 0.2, 0.5, and 1.0 mm. The space-resolved EUV spectrometer employs a holographic laminar-type VLS grating, with an average line spacing of 1200 grooves mm^{-1} (26 mm in groove length and 46 mm in groove distance). The radius of the grating curvature is also a key element in the spectrometer design, which is here 5606 mm. The large curvature of the grating can provide a straightforward reflection of the EUV emission from the entire region of the grating. Figure 3.9 shows typical

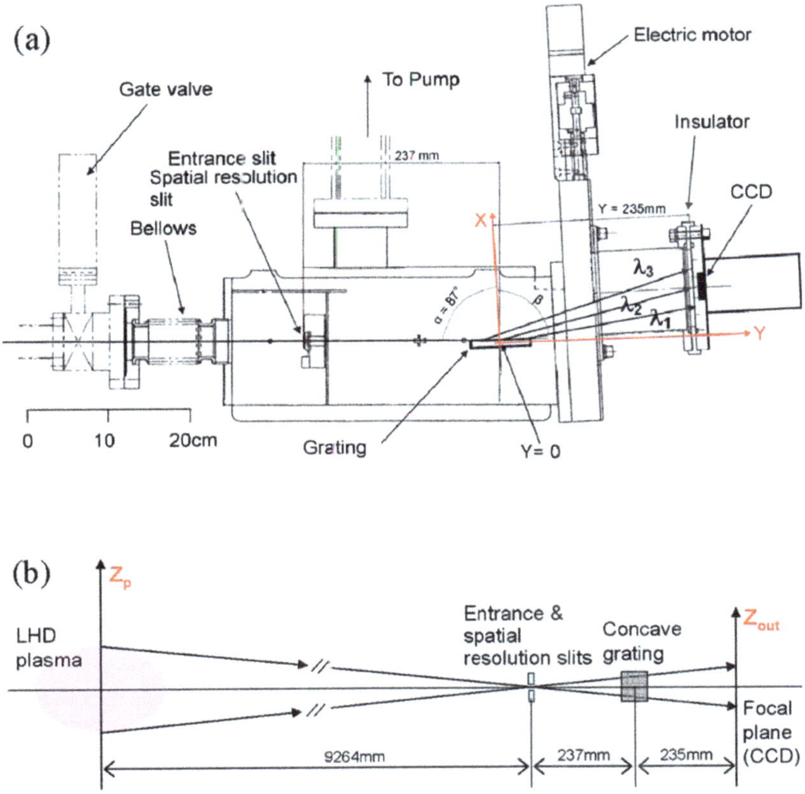

Figure 3.8. Schematic drawings of (a) space-resolved EUV spectrometer in a top view, and (b) principle of spatial resolution in a vertical view. Magnetic surfaces are traced at a plasma axis position of R_{ax} = 3.75 m. The spatial slit for adjusting spatial resolution in the vertical direction is placed on the entrance slit to adjust the wavelength resolution in the horizontal plane. Reprinted from [13], with the permission of AIP Publishing.

vertical profiles of the line emissions from carbon and iron ions observed with the space-resolved EUV spectrometer in the LHD heliotron. As can be seen from figures 3.9(a)–(c), the local maximum location of carbon-line emission around the plasma edge slightly moves toward the plasma center (Z = 0 cm) when the charge state of the carbon ions becomes higher. As shown in figures 3.9(d)–(f), this trend becomes more noticeable in the iron-line case. This experimental result clearly demonstrates the capability of the space-resolved EUV spectrometer in the LHD heliotron.

3.2.2.3 Filtered VUV and EUV diagnostics

As described previously, in contrast to UV–visible–NIR spectroscopy, there is much difficulty in employing filtered diagnostics using such a narrow bandpass filter to measure emissions in the VUV/EUV. However, there is a demand for more easy-to-use and compact filtered diagnostics that can measure VUV/EUV emissions. Such diagnostics could be arranged flexibly to obtain a high throughput and a good

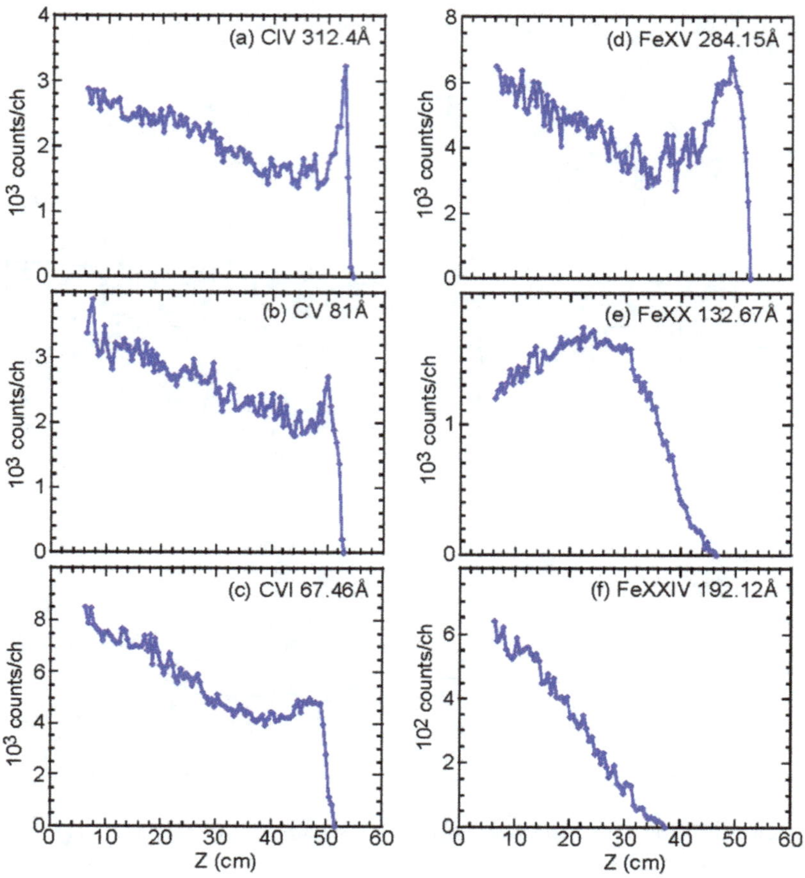

Figure 3.9. Examples of carbon and iron vertical profiles observed in the LHD: (a) C IV (312.4 Å), (b) the second order of C V (2 × 40.7 Å), (c) the second order of C VI (2 × 33.74 Å), (d) Fe XV (284.15 Å), (e) Fe XX (132.67 Å), and (f) Fe XXIV (192.12 Å). Data taken with a sampling time of 200 ms and a binning size of $\Delta X = 5$ and $\Delta Z = 10$. Reprinted from [13], with the permission of AIP Publishing.

spatio-temporal resolution, although a poor spectral resolution is expected. Here, we show several examples of the filtered VUV/EUV diagnostics that have recently been developed to meet such needs.

Some bandpass filters for VUV light are commercially available. The VUV filters are made of fluoride glasses, which are transparent for the light with a wavelength of above about 100 nm. Some important line emissions exist in the VUV domain with a wavelength of above about 100 nm, such as resonance lines of hydrogen (H Lyman-α, 121.6 nm) and carbon (C IV, 155 nm). In MCF experimental devices, the H Lyman-α and C IV lines are usually responsible for radiative power loss from the MCF plasma. Therefore, measurement using PDAs equipped with such VUV filters for H Lyman-α and C IV has been developed [14]. Here, the photodiode used is known as an absolute extreme ultraviolet photodiode (abbreviated as AXUVD). Unlike traditional photodiodes, there is no dead layer in an AXUVD, which makes

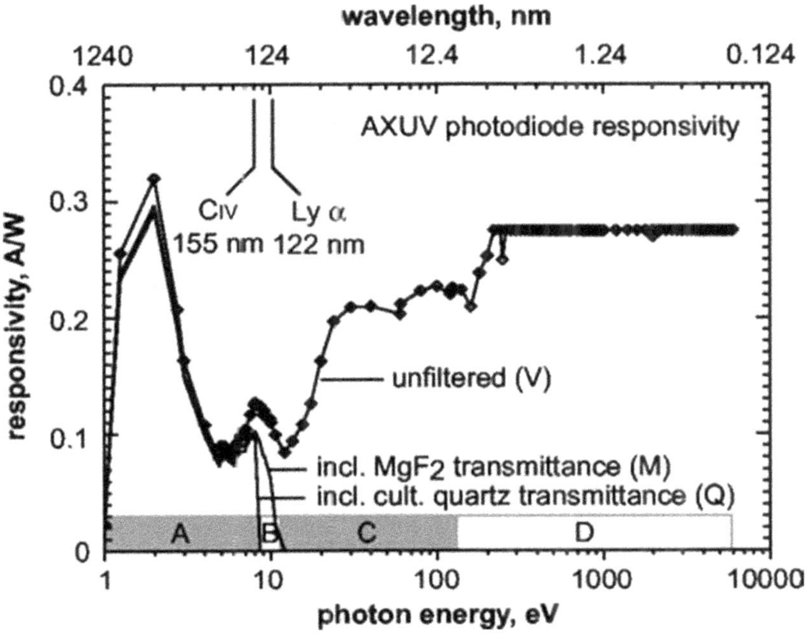

Figure 3.10. Responsivity curves of the AXUVD diagnostics used in the DIII-D tokamak. Curves including the transmittance of the MgF_2 and cultured quartz filters are also included. Curves are based on data from manufacturers. The four energy bands referred to in the discussion are indicated. Reprinted from [14], with the permission of AIP Publishing.

the AXUVD sensitive to VUV and visible light. Although the AXUVD design has several advantages over traditional photodiodes, it should be noted here that the AXUVD has a strong wavelength dependence on sensitivity. Figure 3.10 shows responsivity curves of an AXUVD and such curves considering the transmittance of VUV filters, as utilized in the DIII-D tokamak [14]. As can be seen in the figure, the responsivity of the AXUVD shows a non-uniform curve as a function of the wavelength, and a strong degradation in the AXUVD's responsivity exists at a wavelength of around 124 nm, where the H Lyman-α and C IV exist. In the DIII-D tokamak, in order to extract more accurate information on H Lyman-α and C IV, blind and unfiltered channels are also utilized.

In VUV/EUV filtered diagnostics for light with a wavelength below 100 nm, a method utilizing a Ross filter [15] can be applied. The basic principle of a Ross filter relies on choosing filters with absorption edges located just above and below the energy (i.e., wavelength) of interest. By using a Ross filter, Suzuki *et al* developed filtered diagnostics for a spatial profile of the O IV (55.5 nm, 62.5 nm) and O V (76.0 nm) lines, which are predominant emission lines in the wavelength region from 10–110 nm in the neutral beam injection (NBI)-heated plasmas of the CHS heliotron/torsatron. Here, the Ross filter pair consists of an aluminum (Al) foil (0.2 μm) and a multi-layered foil composed of Al (0.15 μm), LiF (0.05 μm), and Parylene N (0.1 μm).

An alternative to traditional diffraction gratings for VUV/EUV emissions is a multi-layer mirror (MLM), which is made by alternately stacking thin foils with a different refractive index on a substrate. As with Bragg reflection, the reflected light from the interface of each layer is strengthened by the interference. The Bragg equation for the diffracted light to follow can be written as

$$m\lambda = 2d \sin\theta, \tag{3.6}$$

where m, d, and θ are the order of refraction, the thicknesses of the bilayers, and the Bragg angle measured from the surface, respectively. Thus, the MLM acts in the same way as the crystals used in X-ray spectral analysis. Since the necessary thickness is about several tens of nanometers or less for each bilayer, only recent advances in both extremely-thin-film deposition and ultra-smooth substrate polishing technologies have made MLMs practically possible. By increasing the number of bilayers, the reflectivity of the MLM will be increased. However, in contrast to a high-reflection coating for visible light using a similar principle, a 100% reflectivity is not possible for MLMs because there is a certain absorption of light in the VUV to X-ray domains in the bilayer materials. Nonetheless, a reflectivity of 70% or more for EUV light in the vicinity of 13.5 nm has been recently achieved by using a molybdenum (Mo)/silicon (Si) MLM. Owing to such high reflectivity, a high-throughput monochromator can be realized by using MLMs.

Thanks to the compactness of plane MLM monochromators, the Johns Hopkins University Spectroscopy Group developed compact plane MLM monochromators arranged in a fan shape with a pinhole as the entrance slit for the CDX-U and NSTX spherical tokamaks. In addition, they developed a plane MLM high-throughput monochromator, used on a trial basis in the LHD heliotron [16]. In order to enhance the throughput of the monochromator, a grid collimator was employed to collimate the flux of incident light. Figure 3.11 shows the layout of the plane-MLM-based high-throughput monochromator as used on the LHD heliotron. This monochromator is designed for CXS (also referred to as CXRS) in the EUV domain. The impurity ions (magnesium, Mg, ions in this case) in the charge-exchange recombination reaction are injected into the LHD's plasmas by using a tiny plastic pellet embedded with impurities, called a tracer-encapsulated solid pellet (TESPEL), for the purposes of studying the impurity transport. The available amount of the impurities injected by a TESPEL is much less than that of the intrinsic impurity, such as that of carbon, which is usually used for CXS in the visible domain. This problem can be solved when the target domain of the CXS is shifted from the visible to the EUV/SXR domains, since the charge-exchange excitation rates and brightness of a low-n transition in the EUV/SXR domains are usually much higher than those of high-n transitions in the visible domain. Even if that is the case, a high-throughput monochromator is highly necessary to acquire a good time resolution. The grid collimator consists of a number of vertically stacked molybdenum grids, which can provide high throughput (2.5×10^{-4} cm ^{2} sr) with a good radial resolution (<2 cm) on the LHD's equatorial plane. The incident light through the grid collimator is filtered by a Ni/C MLM, which has 2d <120 Å and N <80 reflecting layers, on a

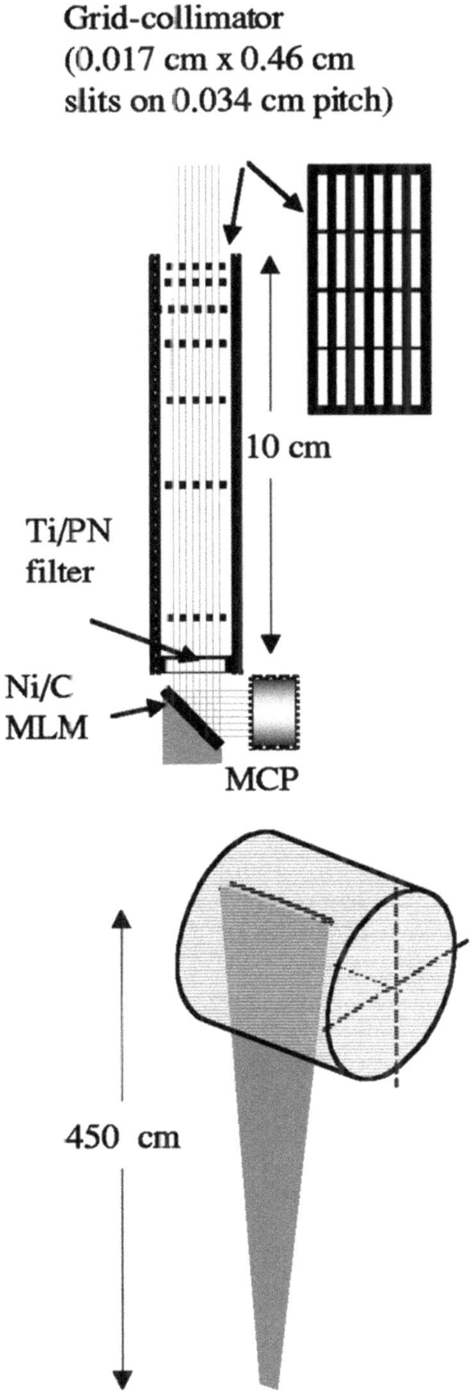

Figure 3.11. Layout of a plane MLM-based high-throughput monochromator, as used in the LHD. Also shown is an instrument view on the LHD. Reprinted from [16], with the permission of AIP Publishing.

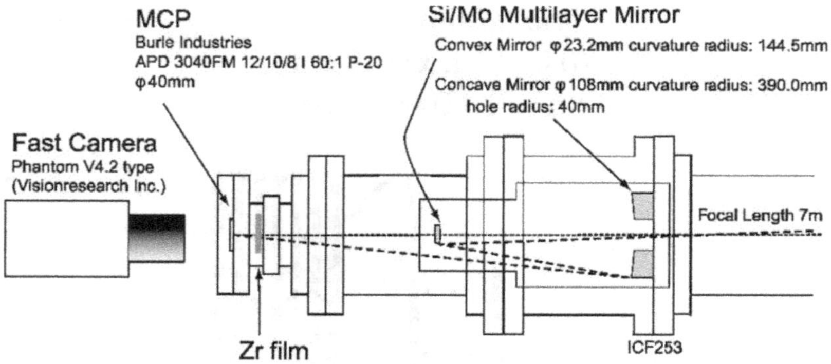

Figure 3.12. Schematic diagram of the tangentially viewing VUV telescope system used in the LHD. Reprinted from [17], with the permission of AIP Publishing.

precisely machined prismatic block. Then, the light with a spectral bandpass of a few angstrom centered around 45.5 Å, which corresponds to Mg Balmer-α, goes to a CsI-coated MCP detector, which is operated in a current mode.

Utilizing concave/convex-type MLMs, an MLM-based telescope has been developed [17]. Figure 3.12 shows a schematic drawing of the developed MLM-based telescope. It employs an inverse-Schwarzschild-type optical layout, consisting of two Mo/Si MLMs for detecting C VI emission (13.5 nm). The characteristics of the MLMs used in this design are a bilayer typical thickness of 6.66 nm and a Mo/Si period ratio of 4:6. A zirconium (Zr) foil with a thickness of 200 nm is inserted to filter out the lower-energy photons (above 20 nm). Thus, the MCP will catch the images, which are composed of photons with about 13.5 nm, and the phosphor screen will convert the EUV images to visible images. One of the unique characteristics of this MLM-based telescope is its utilization of a fast CMOS camera (up to 25 400 images per second with a size of 128 × 128 pixels) as a final detector.

3.2.3 Soft X-ray diagnostics

Plasmas with an electron temperature above 1 keV also emit SXR radiation, as described in the previous subsection. Typical line emissions in the SXR radiation domain from MCF plasmas are K-shell line emissions from medium-Z impurities (e.g., Cr, Fe, and Ni) derived from stainless steel, which is widely used as a construction material in MCF devices. In addition, as described in section 3.1, when a plasma is contaminated by impurities, it is prominently reflected in the level of continuum radiation, due to the Z^4 dependence of the recombination radiation, which is one of the main elements for the continuum radiation. Therefore, measurements of filtered SXR radiation as well as non-filtered, broadband SXR radiation are helpful in the study of impurity concentration and transport in MCF plasmas. Since the bremsstrahlung radiation has a strong dependence on electron density (n_e^2), as given by equation (3.2), moreover, broadband SXR radiation is also useful in the study of the formation and deformation of the confined plasma profiles. In terms of the deformation of the plasma profiles, if broadband SXR measurements with a sufficiently high time resolution are available, such diagnostics can identify the MHD modes structure.

Even for SXR measurements, grazing-incidence spectrometers can still be used with a moderate resolving power, $\lambda/\Delta\lambda$, of about 100–200 (and 500–600 at most). In this subsection, other typical diagnostics for SXR measurements from the broad band to the spectrally high resolved will be explained.

3.2.3.1 Diode array diagnostics

Semiconductor detectors have been widely used in broadband measurements of SXR radiation in MCF plasmas. Currently, silicon PN-junction or PIN diodes are commonly used in various MCF devices. Here, 'I' denotes an intrinsic semiconductor without any dopants; 'P' denotes a P-type semiconductor, which is made by doping an intrinsic semiconductor with an electron acceptor; and 'N' denotes an N-type semiconductor, which is made by doping an intrinsic semiconductor with an electron donor. Thus, a 'PIN' diode is a diode with a wide intrinsic semiconductor region (almost corresponding to the depletion region) between the P-type and N-type semiconductor regions. A PN-junction diode has no intrinsic semiconductor region, but at the junction area the electrons diffused from the N-type semiconductor to the P-type semiconductor come into contact with the holes diffused from the P-type semiconductor to the N-type semiconductor, where they are recombined. Accordingly, the electric charge carriers are depleted in the region around the junction area of PN diodes. In the depletion region, an electric field is naturally generated. In the case of PIN diodes, a reverse bias voltage (e.g., around 15 V is enough) acting in the direction of broadening the depletion region is usually applied due to the existence of the intrinsic semiconductor region. The reserve bias voltage also enhances the electric field in the depletion region. When the photodiode receives the photons, electron–hole pairs are created in the depletion region by the energy, which are released by the incident photons in the depletion region. Although the bandgap energy for silicon is about 1.12 eV, the average energy for the creation of an electron–hole pair is about 3.66 eV. The electric charge carriers generated by the incoming photons are collected sufficiently by the electric field in the depletion region, and the photodiode produces an electric signal. AXUVDs, which as detailed previously have been recently developed, are also used for broadband SXR measurement in MCF experiments. These AXUVDs have the advantage of no doped dead region, which eliminates surface recombination of the electric charge carriers. Thus, AXUVDs have 100% efficiency in the collection of electron charge carriers, and correspondingly demonstrate near-theoretical quantum efficiency for EUV photons and low-energy particles.

A diode has several advantages over other optical detectors, such as PMTs. For instance, there is an excellent linear response over a wide range of photon fluxes. In the current mode, where changes in the diode current are measured, the bandwidth of signals from AXUVDs can be up to several hundreds of KHz and more, although this bandwidth is inferior as compared with PMTs. Silicon photodiodes have a sensitivity of up to 1100 nm. Diodes made with other semiconductor materials can exhibit a sensitivities for longer wavelengths. However, the active area of the diode is usually coated with either silicon nitride, silicon monoxide, or silicon dioxide for the purposes of protection and anti-reflection. Thus, in the case of visible-light measurement, these coatings should be optimized according to the target wavelength of the radiation. In the case of SXR measurement, in order to cut off the lower-energy light, a thin beryllium (Be) foil is installed in between the plasma and diodes. The beryllium foil also can act as a vacuum interface between the vacuum vessel of the MCF devices and the diode detector unit. The thickness of the beryllium foil is determined by the cutoff energy, and typically ranges from several tens to several hundreds of microns.

A single-element diode, as well as multiple-element diode array, can be fabricated compactly at low cost. For direct broadband SXR measurement in MCF devices, an array consisting of 16 or 20 diode elements (i.e., channels) is typically used (sometimes the edge element is blinded for dark measurement). In order to obtain a given spatial resolution with such a diode array measurement, a pinhole or slot aperture is placed between the MCF plasma and the diode array. Figure 3.13 shows layouts of the SXR diode arrays installed in two MCF devices, the W7-X stellarator and the DIII-D tokamak. The SXR diode arrays are usually installed as close as possible to the plasma in order to acquire as wide a view of the poloidal cross-section of the plasma as possible. Consequently, some problems may occur. The radiation heat from the plasma might heat up the SXR diode array, which can increase noise

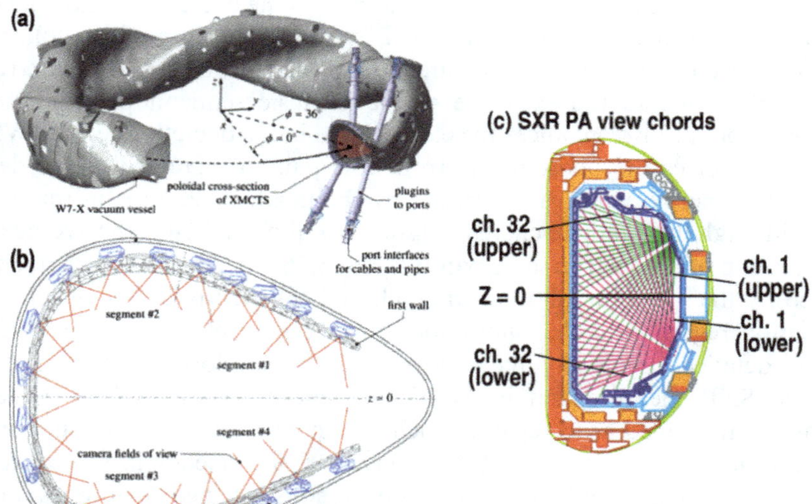

Figure 3.13. Schematic view of the layout of SXR arrays installed in (a) and (b) the W7-X stellarator, and in (c) the DIII-D tokamak. For the W7-X stellarator, 20 × 18-channel SXR arrays are installed inside the vacuum vessel at a toroidal angle of $\phi = 36°$, where the cross-section is triangular in shape (shown in (a)). The fields of view of each array are indicated by the red lines. For the DIII-D tokamak, two sets of 2 × 16-channel SXR arrays are installed at the same toroidal section ($\phi = 90°$). Each array consists of two photodiode arrays viewing slight up/down. Panels (a) and (b) reproduced from [18]. © The Author(s). Published by IOP Publishing Ltd. CC BY 3.0. Panel (c) reprinted from [19], with the permission of AIP Publishing.

in the diode current. When the increment in temperature of the SXR diode array becomes a problem, a water-cooling circulation system is usually built into the SXR diode array unit. In MCF devices, glow discharge cleaning is usually implemented to condition the PFCs in the vacuum vessel. After glow discharge cleaning, a thin film is deposited on the surface of the PFCs in the vacuum vessel, which also might cover the surface of the SXR diode array. Thus, a shutter mechanism to close the aperture window is also installed in SXR diode array units to avoid such surface contamination by the glow discharge cleaning process.

In either device, in order to perform a 2D tomographic reconstruction of the SXR emissivity from a plasma, the SXR diode arrays are installed at as many different poloidal angles as possible on the same poloidal cross-section. However, in most MCF devices a full-angle coverage of the poloidal cross-section of the confined plasma is impossible due to the inner structure of the MCF devices (even in the example case shown in figure 3.13(b), views from some poloidal angles are missing). Rather, in the 2D tomographic reconstruction, the local emissivity is estimated from the sight-line-integrated data. Such an inversion of the integral equation, especially in the case of MCF plasmas, can be categorized as an ill-posed problem. In order to solve the ill-posed problem, many mathematical methods, such as regularization, have been developed and new approaches, such as a compressed sensing, are still being developed. Detailed explanation of such mathematical methods for 2D tomographic reconstruction is beyond the scope of this book; more detailed information on tomography in MCF plasmas can be found in the review paper by Ingesson *et al* [20]. Figure 3.14 shows an example of 2D SXR tomographic images based on the SXR array diagnostics in the JET tokamak [21]. In the JET tokamak, in order to examine the impact of ion cyclotron resonance heating (ICRH) on the impurity behavior, a comparison experiment between NBI and NBI + ICRH was performed. Figure 3.14(a) was obtained for the plasma with NBI only, and figure 3.14(b) was obtained for the plasma with NBI + ICRH. As can be easily recognized, the 2D SXR radiation profile shows an intense spot in the plasma center in the case of NBI while such a spot has disappeared in the radiation profile for the NBI + ICRH case. Instead, the SXR emissivity from the outer region at $R \sim 3.6$ m is increased. Therefore, this comparison experiment clearly suggests that the impurity ions which accumulate in the plasma center move toward the outer region when applying ICRH. Thus, by utilizing SXR tomographic image reconstruction, 2D representations of the behavior of impurities in MCF plasmas can be obtained.

3.2.3.2 Pulse height analyzers

A moderate energy-resolved measurement can be performed using a pulse height analyzer (PHA) with a photon-counting detector. In general, PHA systems also utilize a semiconductor detector. The number of electron–hole pairs that are created by an incident X-ray in the depletion region of a semiconductor is proportional to the energy of the X-ray photon. In other words, the resulting amplitude (i.e., pulse height) of the output voltage pulse from the semiconductor detector is proportional to the energy of the X-ray photon received by the semiconductor detector. By

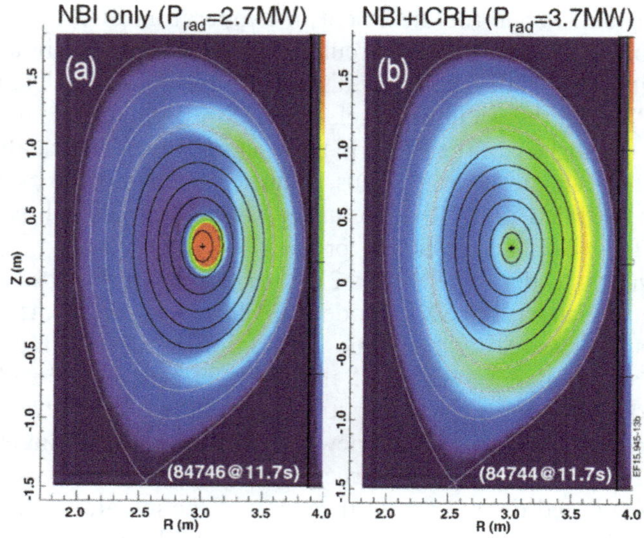

Figure 3.14. Examples of two-dimensional SXR tomography: comparison of two discharges, one with NBI and one with NBI + ICRH in JET tokamak. The color bar is the same for the two images. An intense spot of SXR radiation can be seen in the plasma center in the case of NBI. Reproduced from [21]

applying the pulse height analysis technique on the amplitudes of the voltage pulse signals within a given time interval (typically 10 ms to 100 ms), the temporal evolution of the X-ray spectra from the MCF plasmas can be obtained. Lithium-drifted silicon, Si(Li), diodes are conventionally used as the detector in PHA systems. Such Si(Li) diodes have a wide depletion region, so that X-ray photons with an energy of up to 30 keV can be detected. In order to reduce the dark-current noise generated in the wide depletion region, the Si(Li) diodes must be cooled down by liquid nitrogen. A typical energy resolution of 230 eV FWHM at 5.9 keV with the Si(Li) PHA system is sufficient to observe closely appeared K-shell line emissions from medium-Z impurities separately. For X-rays with higher energies of up to 100 keV and more, a high-purity germanium (or HPGe) detector is typically used. Recently, a silicon-drift detector (SDD) has been developed. This SDD can be operated at higher count rates with a similar energy resolution as the conventional Si (Li) detector. Thus, a PHA system utilizing a SDD can be operated with a higher temporal resolution compared with conventional Si(Li) PHA systems. A SDD achieves this performance by cooling with a Peltier device rather than with liquid nitrogen. In this way, a SDD can also help to miniaturize the PHA system.

Figure 3.15 shows typical SXR energy spectra measured with the PHA system installed on the TFTR tokamak [22]. The PHA system in the TFTR tokamak consisted of six PHA detectors with different filters and observation geometries. The upper trace of figure 3.15 was made of corrected data from three of the six PHA detectors. The lower trace of figure 3.15 was synthesized by the data from all of the six PHA detectors. Even from the upper trace, Kα emissions from representative

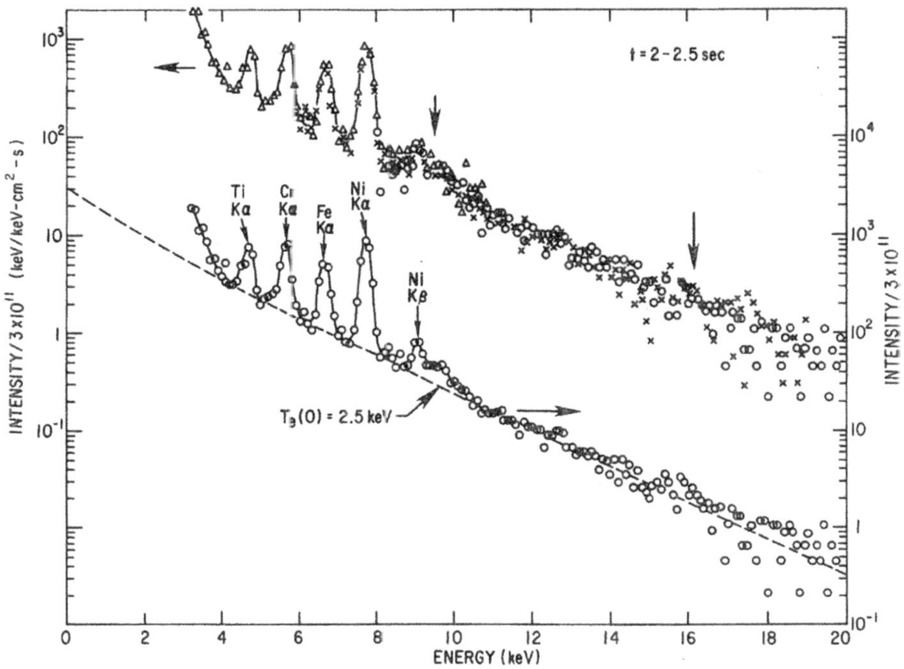

Figure 3.15. SXR energy spectra measured by the PHA system on the TFTR tokamak. The upper trace consisted of corrected data from three of the six PHA detectors with different filters and observation geometries. The lower trace was synthesized from the six PHA detectors. Reprinted from [22], with the permission of AIP Publishing.

medium-Z impurities, such as titanium, Cr, Fe, and Ni, are clearly seen. From the synthesized energy spectra, Ni Kβ emissions can also be seen. The electron temperature, which can be deduced from the slope of the SXR energy spectra, is estimated at 2.5 keV.

3.2.3.3 X-ray crystal spectroscopy

A crystal spectrometer can measure line emissions from impurity ions in the SXR domain with a high spectral resolution. The primary purpose of a crystal spectrometer is to measure the ion temperature from the Doppler broadening of the line emission and the toroidal flow velocity from the Doppler shift of the line emission. To this end, the helium (He)-like $1s^2\ ^1S_0$–$1s2p\ ^1P_1$ resonance line profile of argon (Ar) or iron ions is usually measured, because the helium-like ions of argon or iron can exist over a wide range (1–6 keV) of electron temperature. When the central electron temperature exceeds 10 keV, krypton will be the best option for measurement. It should be noted here that the ion temperature obtained by the crystal spectrometer is the temperature of the impurity ions which are indigenous to the plasma or are introduced artificially from outside, not exactly the temperature of the bulk ions, such as hydrogen isotopes or helium. When the helium-like resonance line of different impurity ions needs to be measured, different crystals set at the

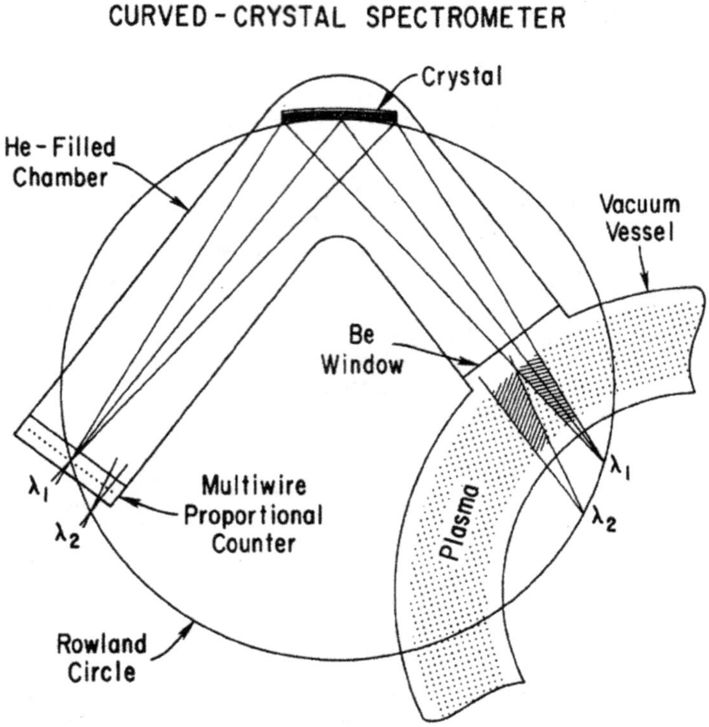

Figure 3.16. Schematic view of the setup of the curved-crystal spectrometer with a Johann configuration used on the PLT tokamak. Reprinted with permission from [23], Copyright (1979) by the American Physical Society.

corresponding Bragg condition should be used. For instance, in the WEST (Tore Supra) tokamak, to measure the helium-like and the lithium-like ions of 10 impurities between argon and krypton, eight crystals in total are mounted on the surface of a rotary barrel, which can be selected remotely between discharges, in the spectrometer used (one crystal works for several impurities). By using a crystal spectrometer, the electron temperature and charge-state distribution of the impurity ions of interest can also be deduced from the line-intensity ratio of the helium-like resonance line to the dielectric and inner-shell excited satellite lines from lithium-like ions. Of course, from the intensity of the line emissions from the impurity ions, the impurity ion density can also be obtained. From these points of view, a crystal spectrometer is recognized as a powerful tool in the study of impurity transport in high-temperature MCF plasmas.

There are several optical configurations for a crystal spectrometer. In MCF experiments, Johann-type crystal spectrometers are most widely used. Figure 3.16 shows a schematic drawing of a curved-crystal spectrometer as used on the PLT tokamak [23]. A germanium (220) crystal, which was bent by a specially made bending jig to a radius of curvature of 2.42 m, was originally used in this crystal spectrometer. By replacing the germanium crystal with a quartz crystal with a radius of curvature of 3.33 m, the energy resolution of the instrument was improved from 1700 to 15 000. As the photon detector, a position-sensitive multi-wire proportional counter (MWPC) was used in this crystal spectrometer. Consequently, temporally resolved X-ray spectra were obtained (in this case, a temporal resolution of about 150 ms as achieved). In recent conventional crystal spectrometers, a CCD detector dedicated to X-ray measurement is widely used as the photon detector. A typical argon spectrum measured with the crystal spectrometer equipped with a MWPC detector on the Tokamak Experiment for Technology Oriented Research (TEXTOR) is shown in figure 3.17 [24]. The lines shown here are labeled according to the widely used nomenclature of Gabriel. Line W is the helium-like $1s^2\ ^1S_0$–$1s2p\ ^1P_1$ resonance line, which is commonly used for ion temperature and flow velocity measurements. Lines X and Y are the helium-like intercombination lines of $1s^2\ ^1S_0$–$1s2p\ ^3P_2$ and $1s2p\ ^3P_1$, line Z is the helium-like $1s^2\ ^1S_0$–$1s2s\ ^3S_1$ forbidden line, and lines j, k, and q are the associated lithium-like satellite lines ($1s^2\ nl$–$1s2l'nl''$ transitions with $n = 2$). As described above, the electron temperature can be deduced from the line-intensity ratio of j/w (or k/w) and the ratio of the lithium-like and helium-like ions can be given from the ratio of q/w.

In contrast, the crystal spectrometers installed in the Alcator C-MOD tokamak employ a von Hamos configuration. A crystal spectrometer with a von Hamos configuration usually has a lower throughput compared to that with a Johann configuration, due to the use of an entrance slit. Further, in a von Hamos-type crystal spectrometer, light of a certain wavelength is reflected on one part of the crystal surface, not on the entire crystal surface. This also contributes to aggravating the throughput of the crystal spectrometer. The reasons why the Alcator C-MOD tokamak employs a crystal spectrometer with a von Hamos configuration are as follows. The expected SXR/X-ray emission intensity from the Alcator C-MOD plasmas is higher than that from other tokamaks because the SXR/X-ray emission intensity is generally proportional to the square of the electron density, and the typical electron density of the Alcator C-MOD plasmas is higher than that in other tokamaks. Moreover, the von Hamos-type crystal spectrometer is normally more compact than the Johann-type crystal spectrometer under the same spectral resolution. By taking advantage of this compactness, the Alcator C-MOD tokamak arrays five von Hamos-type crystal spectrometers with different lines of sight in order to measure the spatial profile of the ion temperature and the toroidal flow velocity.

Recently, in order to simplify measurement of the spatial profile of the ion temperature and plasma flow velocity, a spatially resolved crystal spectrometer has been developed. Such a spectrometer, developed by the Princeton Plasma Physics Laboratory, is known as an X-ray imaging crystal spectrometer (XICS). An XICS can also provide information on the spatial profile of the density of highly ionized impurity ions. As an photon detector, an X-ray CCD detector can of course be used, but the readout speed of the CCD image becomes very slow when acquiring the X-ray spectra with sufficient spatial and spectral resolutions, even by binning pixels in the X-ray CCD detector. In order to obtain high spatially and spectrally resolved X-ray spectra with a good temporal resolution, a hybrid pixel detector is utilized as

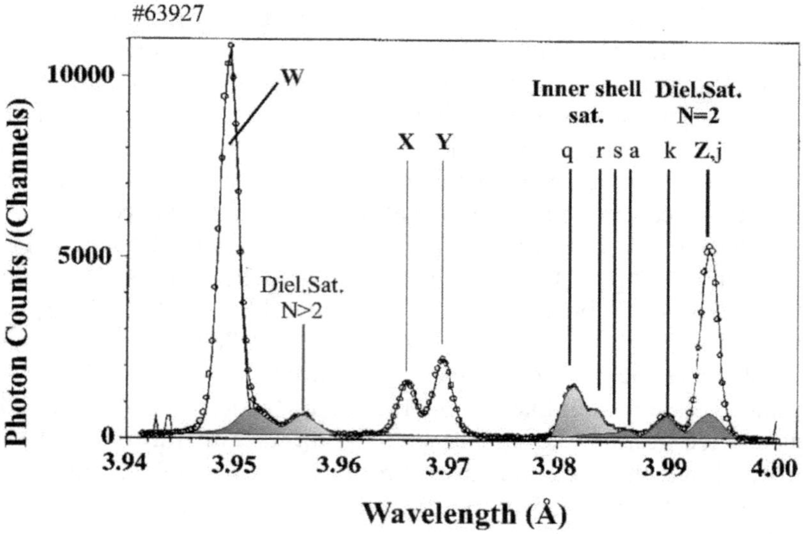

Key	Transition	Wavelength (Å)*
W	$\mathbf{1s^2\,{}^1S_0 - 1s2p\,{}^1P_1}$	3.9494
X	$\mathbf{1s^2\,{}^1S_0 - 1s2p\,{}^3P_2}$	3.9661
Y	$\mathbf{1s^2\,{}^1S_0 - 1s2p\,{}^3P_1}$	3.9696
q	$1s^22s\,{}^2S_{1/2} - 1s2s2p\,({}^1P)\,{}^2P_{3/2}$	3.9815
r	$1s^22s\,{}^2S_{1/2} - 1s2s2p\,({}^1P)\,{}^2P_{1/2}$	3.9836
a	$1s^22p\,{}^2P_{3/2} - 1s2p^2\,{}^2P_{3/2}$	3.9860
k	$1s^22p\,{}^2P_{1/2} - 1s2p^2\,{}^2D_{3/2}$	3.9901
j	$1s^22p\,{}^2P_{3/2} - 1s2p^2\,{}^2D_{5/2}$	3.9941
Z	$\mathbf{1s^2\,{}^1S_0 - 1s2s\,{}^3S_1}$	3.9944

* Theory by Vainshtein and Safranova

Figure 3.17. Experimental Ar XVII spectrum with corresponding satellites obtained on TEXTOR-94. Lines originating from Ar XVI and Ar XVIII are also present. The shaded regions illustrate the contributions from the dielectronic satellites to the spectrum. Reprinted from [24], with the permission of AIP Publishing.

the photon detector of the XICS. A PILATUS detector is one such well-known hybrid pixel detector, originally developed at the Paul Scherrer Institute at the Swiss Light Source for the experiment at CERN. PILATUS detectors are characterized by a much larger dynamic range and a higher maximum count rate compared to conventional MWPC or X-ray CCD detectors. Therefore, the XICS is a very

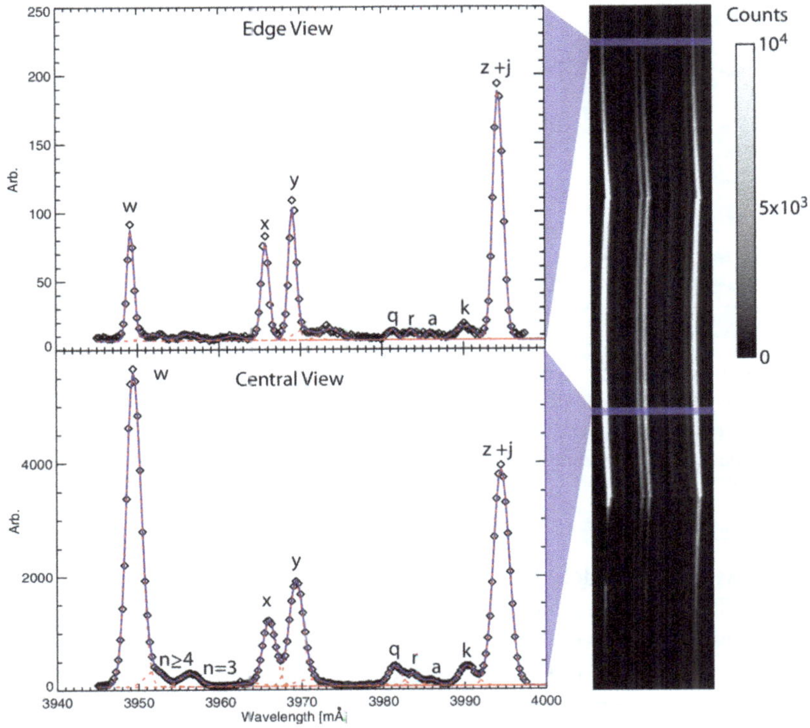

Figure 3.18. Chord-integrated helium-like argon spectra and satellites from two sight lines through Alcator C-Mod plasma (left) and raw images from the three detector modules viewing the He-like spectra (right). The total multi-line fits are shown in solid blue, while the intensities associated with individual lines are shown in dashed red. The greater widths of the lines in the central-view spectrum compared to those in the edge-view spectrum reflect the difference in ion temperature between these two parts of the plasma. Reprinted from [25], with the permission of AIP Publishing.

powerful tool for the study of impurity transport in high-temperature MCF plasmas. Figure 3.18 shows typical argon spectra measured with the XICS, equipped with PILATUS detectors, installed on the Alcator C-MOD tokamak [25]. The right figure consists of raw images around helium-like spectra taken by the three detector modules. The upper region comes from the plasma edge and the central region comes from the plasma center. The left figures show the sight-line-integrated, helium-like argon-line emissions and satellite line emissions from two distinct regions, the plasma edge and the plasma center. As can be clearly seen in the left figures, though similar spectra are obtained the widths of the spectral lines from the plasma center are found to be larger than those from the plasma edge. Therefore, although the XICS measures sight-line-integrated spectra, the difference in ion temperature between two regions of the plasmas can be obtained. Some inversion techniques should be applied to the XICS data to estimate the spatial profiles of ion temperature, toroidal flow velocity, and impurity ion density.

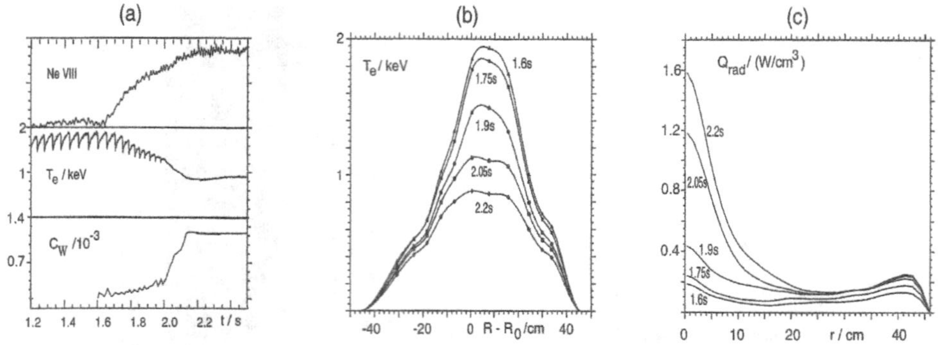

Figure 3.19. Example of impurity accumulation obtained in the TEXTOR tokamak plasma. Shown are the temporal evolutions of Ne viii radiation, central electron temperature and tungsten concentration (left), and the evolutions of the electron temperature profile (middle) and the radiated power density (right). Reproduced from [26] © International Atomic Energy Agency. Published by IOP Publishing. All rights reserved.

3.2.4 Bolometers

Bolometers have been widely utilized for the measurement of plasma radiation in almost all MCF devices. Bolometers are primarily used for the study of power balance analysis, which provides information regarding how much power input into the plasma is confined in the plasma by subtracting the power losses such as the plasma radiation. In a steady state, by solving the balance between input and output energies, the heat transport property of the electrons and the ions can be estimated. In principle, the total plasma radiation can be numerically estimated if all the components responsible for the total plasma radiation are known. In this way, the contributions from specific species of impurity to the total plasma radiation can also be estimated. However, in reality, such estimations are almost impossible, because usually it is difficult to measure the 'absolute' total plasma radiation. Further, the procedure of considering all the radiation components in exhaustive detail is too impractical to be utilized for such a purpose. Hence, it is very difficult to perform practically a detailed study of the impurity transport with only the information regarding plasma radiation. Nevertheless, when the radiation from the impurities of interest is notably dominant, for example when impurity accumulation occurs in the core region of an MCF plasma, the behavior of the plasma radiation could reflect the behavior of the impurities in the plasma. An example of the impurity accumulation obtained in the TEXTOR tokamak is shown in figure 3.19 [26]. In the TEXTOR plasma, heated additionally with a hydrogen NBI, the accumulation of tungsten has occurred after a neon gas puff. When the concentration of tungsten is increased, the core electron temperature decreases and the core radiated power measured with the bolometer increases rapidly. Thus, the bolometric diagnostic can be used to track notable changes in the spatial distributions of impurities in MCF plasmas.

The detector design used in bolometric measurement consists, in principle, of a heat absorber and a thermometer. The absorber is heated by the plasma radiation, which increase in the temperature of the absorber is monitored by the thermometer.

The increment of temperature in the thermometer is converted to the absorbed radiated power, and finally to the plasma radiation from the MCF plasma. In the case of resistive bolometers, the most widely used type of bolometer in MCF devices, the temperature rise of the heat absorber is measured by the change in resistances divided by a thermal capacity, from which the radiated power from the plasma can be obtained. The absorber determines the spectral response of the bolometer. In order to estimate absolutely the radiated power from the plasma, a 100% sensitivity to the radiation from the infrared to the X-ray domains is desirable. To this end, a thin foil made of metals, such as gold and platinum, is used as a heat absorber. Regarding the temporal response of the bolometers, the thermal conductivity in the bolometer and the signal noise level determines the total temporal resolution. The typical time resolution of currently available bolometers is in the order of 10 ms, which can be reduced to a millisecond order in cases with sufficiently high signal-to-noise ratio. In recent MCF experiments, an unfiltered AXUVD has been also utilized as a bolometer, largely owing to its broadband sensitivity from the visible to SXR domains. However, as already discussed before, since an AXUVD has a non-uniform sensitivity over the wavelength range of emissions from MCF plasmas, there is a recognizable difference in the radiated power measured with foil-resistive bolometers and AXUVDs. Nevertheless, AXUVDs still present an advantage over conventional resistive bolometers, thanks to their low cost, low noise, and high temporal resolution.

In early MCF experiments, single uncollimated bolometers, so-called wide-angle bolometers, were mainly utilized. The estimation of total plasma radiation significantly depends on the plasma radiation distribution in the plasma. If the spatial profile of the plasma radiation deviates from both poloidal and toroidal symmetries (otherwise known as multifaceted asymmetric radiation from the edge, or MARFE for short [27]), the total plasma radiation from the measurement could be overestimated or underestimated. Thus, bolometer arrays have been utilized to measure a one-dimensional poloidal or toroidal distribution of the plasma radiation. In order to measure a poloidally and/or toroidally asymmetric distribution of plasma radiation in MCF plasmas simultaneously, we require a 2D bolometric diagnostic with high time and spatial resolution. In order to obtain such a diagnostic, as a matter of course, the arrangement of the bolometer detector elements in a matrix was first considered. However, it is a significantly burdensome and expensive task to prepare all the bolometer elements needed to achieve a good spatial resolution. To reduce the cost and time of preparing such a 2D bolometric diagnostic, an 'imaging' bolometer has been developed at the National Institute for Fusion Science (NIFS), Japan. Recently, an improved version of such an imaging bolometer, an infrared imaging video bolometer (IRVB), also developed at NIFS, has been utilized in several MCF devices, such as the LHD, JAERI Tokamak-60 Upgrade (JT-60U), and KSTAR. An IRVB uses a large-sized thin metal foil in the frame, the temperature rise of which is then measured by an infrared camera. The infrared camera used determines the temporal resolution of the IRVB. This configuration, consisting of the metal foil and the infrared camera, provides users with some latitude in terms of sensitivity, but which has a trade-off with the spatial resolution.

3.3 Active spectroscopy

3.3.1 Principle of charge-exchange spectroscopy

Passive spectroscopy has been widely used as a useful diagnostic of plasma parameters. However, passive spectroscopy does not provide the local quantities (temperature flow velocity and density), because the spectrum measured is always an integrated signal along a line of sight. In order to realize local measurements with spectroscopy, active spectroscopy, in which the emission is locally excited by a laser or neutral beam, is required [28, 29]. Spectroscopy of line emission from impurity ions excited by charge-exchange recombination reactions with neutral beam injection was developed in the 1980s, and has since been used as a common diagnostic for ion temperature, flow velocity, and impurity ion density profiles in magnetically confined plasmas. The charge-exchange recombination reactions between a fully ionized impurity, A^{z+}, with a charge of z, and a hydrogen (H^0) or deuterium (D^0) neutral beam results in an excited ion with one electron, $A^{(z-1)+}$, expressed as

$$H^0(D^0) + A^{z+} \rightarrow H^+(D^+) + A^{(z-1)+}(n) \tag{3.7}$$

$$\rightarrow H^+(D^+) + A^{(z-1)+}(n') + h\nu(\Delta n = n - n'). \tag{3.8}$$

Here, n and n' are the upper and lower levels of the excited state. Spectroscopy in the wavelength ranges that can be transferred via optical fibers is widely used because of the fibers' easy access to nuclear fusion devices. The $\Delta n = 1$ transition gives the emission in the visible range when the upper level is high enough, for example $n = 4$ for He^+ (468.6 nm) and $n = 8$ for C^{5+} (529.2 nm). The local ion temperature, flow velocity, and impurity density are derived from the Doppler width, Doppler shift, and total intensity of the spectra. The total intensity of the charge-exchange lines, B_λ^{CX} is given by [28]

$$B_\lambda^{CX} = \frac{1}{4\pi}\sum_{i=1}^{M}\langle\sigma v\rangle_\lambda^i \int n_z n_b^i dl, \tag{3.9}$$

where $\langle\sigma v\rangle_\lambda^i$ is the emission cross-section for the transition with a wavelength of λ and the ith component of the neutral beam. The neutral beam with a positive ion source has three energy components, with E, $E/2$, and $E/3$, where E is the acceleration voltage. n_z is the density of measured impurities, and n_b^i is the density of each component of the neutral beam with ith order ($i = 1, 2, 3$ for E, $E/2$, $E/3$). dl is the integration along the line of sight within the neutral beam that is not along the plasma radius.

Figure 3.20 illustrates the principle of CXS in measuring the ion temperature and flow velocity. The measurement is restricted to within the intersection region of the neutral beam and the line of sight, which is a significant advantage over passive spectroscopy, where the observed spectra always suffer line integration along the line of sight. The spatial resolution depends on the geometry of the neutral beam, the magnetic field, and the line of sight. Generally, the spatial resolution is better near

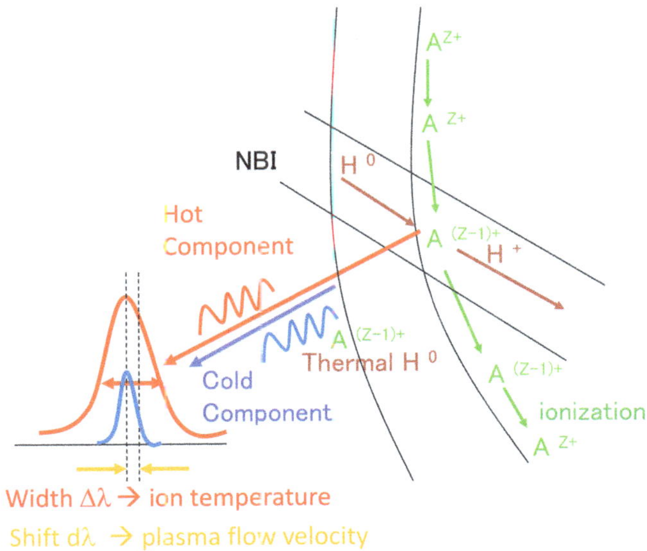

Figure 3.20. Schematic of principle of CXS, used to measure ion temperature and flow velocity.

the plasma edge, where it can be a few millimetres, and where the line of sight is parallel to the magnetic field line. Three parameters are driven by the best fit of the measured spectrum to the Gaussian profile. As aforementioned, the Doppler width and the Doppler shift give the ion temperature and plasma flow velocity.

The first measurement of ion temperature and plasma flow velocity was performed in the Poloidal Divertor Experiment (PDX) tokamak, using the helium line of He^+ (468.6 nm) [28]. A vibrating mirror was implemented to scan the wavelength of the spectrometer and to measure the shape of the spectra of the charge-exchange emission. The spectra of the helium line of He^+ (468.6 nm) at neutral beam-off timing can be fitted with one Gaussian, while the spectra of the helium line of He^+ (468.6 nm) at neutral beam-on timing require two Gaussians to be fitted. The spectra with narrow Doppler width are the background line emission from the plasma edge where the ion temperature is low, while the spectra with wide Doppler width are the emission from the plasma core at the cross-section between the neutral beam and the optical viewing line. Since this first measurement, CXS has been used for most toroidal plasma experiments with neutral beam injection using various lines in the visible region.

The wavelengths of the charge-exchange lines for the transition of various impurities are detailed in table 3.2. The wavelengths become shorter for transitions of higher-Z impurities, while they become longer for transitions at higher n ($\Delta n = 1$). Therefore, a higher-n transition is used for measurement using a higher-Z impurity. In high-temperature toroidal plasma, the selection of lines depends on the material of the wall or of the divertor. Lithium (449.92 nm), beryllium (465.89 nm), boron (494.50 nm), carbon (343.40 nm, 529.09 nm), oxygen (297.61 nm), and neon (524.93

Table 3.2. Table of wavelengths of charge-exchange lines.

λ_0 (nm)	$n = 4 \rightarrow 3$	$n = 5 \rightarrow 4$	$n = 6 \rightarrow 5$	$n = 7 \rightarrow 6$	$n = 8 \rightarrow 7$	$n = 9 \rightarrow 8$	$n = 10 \rightarrow 9$	$n = 11 \rightarrow 10$
He II	**468.61**	1012.4	1863.7	3090.9	4762.2	6946.1	9710.8	13 125
Li III	208.26	**449.92**	828.27	1373.7	2116.4	3087.0	4315.7	5832.9
Be IV	117.14	253.08	**465.89**	772.66	1190.5	1736.4	2427.5	3281.0
B V	74.971	161.97	**298.17**	**494.50**	761.89	1111.3	1553.6	2099.8
C VI	52.063	112.48	207.06	**343.40**	**529.09**	771.72	1078.9	1458.2
N VII	38.250	82.635	152.12	252.29	**388.71**	**566.97**	792.64	1071.3
O VIII	29.285	63.267	116.47	193.16	**297.61**	**434.09**	**606.86**	820.21
F IX	23.139	49.488	92.024	152.62	235.15	**342.98**	**479.49**	**648.06**
Ne X	18.742	40.491	74.539	123.62	190.47	277.81	**388.39**	**524.93**

nm) charge-exchange lines have been used for measurements in high-temperature toroidal plasmas.

3.3.2 Background line emission

Background line emission is caused by the charge-exchange reaction between fully ionized impurities and the thermal neutral at the plasma edge, where the ion temperature is low. Therefore, this background line emission is referred to as the cold component. In contrast, the charge-exchange recombination (CXR) emission due to the charge-exchange reaction between fully ionized impurities and the neutral beam has a wider Doppler width and is referred to as the hot component, because the ion temperature at the intersection between the line of sight and the neutral beam line in the core plasma is higher than at the plasma edge. There is a slight difference in the zero-Doppler-shifted wavelength between the hot component and cold component. The wavelength difference is much smaller than the Doppler width (typically less than 0.1% of the FWHM of the hot component). Therefore, the difference in the observed wavelength between the hot component and the cold component is mainly due to the difference in the Doppler shift between the core and edge plasma. When the plasma flow velocity is small, the separation of hot and cold components becomes difficult due to the small Doppler shift. The cold component causes underestimation of the ion temperature because the charge-exchange line measured is always the sum of the cold and the hot components.

Therefore, subtraction of the cold component is essential in CXS in order to derive the local ion temperature, flow velocity, and density of impurity in the plasma core. Three techniques have been applied to extract the hot component.

(1) *Two-Gaussian fitting.* This technique is applied to the CXS for the first time. This technique is valuable when the difference in the Doppler width or shift of the cold and hot components is large enough to be fitted. However, when the intersection between the neutral beam and the optical viewing approaches the plasma edge, the separation of the two Gaussians (cold and hot components) becomes difficult.

(2) *Two sets of viewing arrays for on-beam line and off-beam line.* This technique measures the cold component and the cold + hot components by installing two sets of optical fiber arrays viewing on-beam line and off-beam line. The hot component can be extracted by subtracting the signal of the off-beam-line viewing array from that of the on-beam line viewing array. The advantage of this technique is that it can be applied to plasma in the transient phase, differing from the technique used in (3). However, a toroidal symmetry of the cold component is assumed in this technique. This assumption is usually satisfied in divertor configurations, but not in limiter configurations because the limiter enhances the thermal neutral and the intensity of the cold component locally and breaks the toroidal symmetry.

(3) *Neutral beam modulation.* This technique requires a special operation in the neutral beam injection. In this, the neutral beam needs to be modulated (i.e.,

to repeat the NBI on and NBI off in time) to measure only the cold component. The spectra of neutral beam-off timing consists of the cold component, while the spectra of neutral beam-on timing consists of the cold and hot components. By subtracting the signal during the neutral beam-off timing from that during neutral beam-on timing, the hot component spectra can be obtained if the intensity of the cold component is unchanged during the on–off modulation of the neutral beam. The duty cycle of the beam modulation (the ratio of beam-on time to one period of modulation) can be flexible (not necessarily 50%) depending on the time-averaged NBI power required for the experiment. A low duty cycle (<10%) is used for measurements in ohmic plasma, while a high duty cycle (80%–99%) is used for high-power NBI-heated plasma. The beam-off time should be longer than the integration time of the measurement, which is determined by the number of photons deemed necessary for accurate measurement. This technique is suitable in the steady-state phase, where the intensity of the cold component is constant in time. However, the technique may encounter problems in the transient phase, where the cold component changes in time. This technique is now the most widely used in experiments.

3.3.3 Fine structure

The energy for a given level for the principal quantum number (n) and total angular momentum quantum number (j) in a hydrogen-like ion with the nuclear charge Z is given by

$$E(n, j) = \frac{Z^2 R_\infty(\mu/m_e)}{n^2}\left[1 + \frac{(\alpha Z)^2}{n}\left(\frac{1}{j + 1/2} - \frac{3}{4n}\right)\right], \tag{3.10}$$

where Z, R_∞, and α are the nuclear charge, the Rydberg constant and the fine-structure constant, known as the Sommerfeld constant, respectively. The μ is the reduced mass, defined as $\mu = m_e A m_p/(m_e + A m_p)$, where m_e and m_p are the mass of the electron and the proton and A is the mass number, respectively. The energy level is determined by the total angular momentum quantum number $j(=l + s)$, where l is the orbital angular momentum and s is the spin angular momentum quantum number. Therefore, two energy levels with different l are degenerate; for example, the energy level of the ${}^2P_{3/2}$ ($l = 1$, $s = 1/2$) state is the same as the energy level for the ${}^2D_{3/2}$ ($l = 2$, $s = -1/2$) state.

Figure 3.21 shows the $\Delta l = 1$ transition diagram between the $n = 8$ and $n = 7$ levels in hydrogen-like carbon (C^{5+}). There are eight l sub-levels for the $n = 8$ level and seven l sub-levels for the $n = 8$ level. Each l level has two levels of $s = +1/2$ and $s = -1/2$. The branching ratio between the discrete s is indicated by the number in the diagram normalized by the transition from $s = -1/2$ for $s = +1/2$. The line intensity for the $\Delta s = 0$ transition (from $s = +1/2$ to $s = +1/2$ and from $s = -1/2$ to $s = -1/2$) is much larger than the intensity for the $\Delta s = 1$ transition (from $s = -1/2$

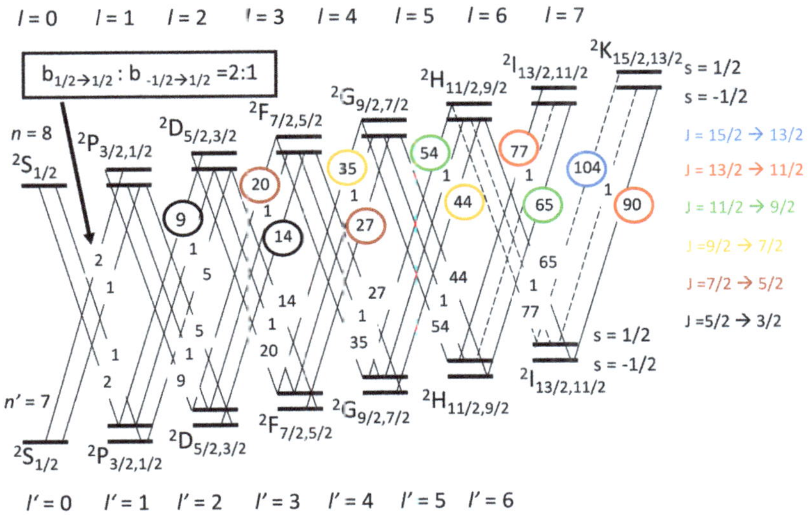

Figure 3.21. Transition diagram between n = 8 and n = 7 levels in hydrogen-like carbon.

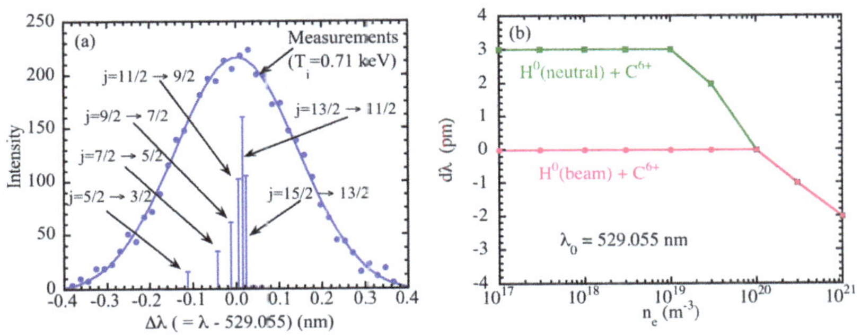

Figure 3.22. (a) Spectra of the n = 8–7 transition of hydrogen-like carbon in a plasma with an ion temperature of 0.71 keV, and (b) density dependence of averaged wavelength of charge-exchange line by the reaction with thermal neutral and beam neutral.

to $s = +1/2$). The transitions for the different $\Delta j = 1$ transitions are represented by different colors.

Figure 3.22 shows an example of charge-exchange spectra from a plasma core during neutral beam injection. Here, the wavelength for the main six $\Delta j = 1$ transitions with separate j levels of $n = 8$ to $n' = 7$ of C^{5+} are plotted. The solid lines are the best-fitted curves using six Gaussian profiles with the amplitude proportional to the intensity ratio and with the same Doppler width. The most intense line is the transition from $j = 13/2$ to $j = 11/2$. One transition is in the redshifted side and the other four lines are at the blueshifted side. Because the Doppler width is much larger than the separation of wavelength with different j transition, the spectra in the plasma core are almost symmetric and close to a single Gaussian profile. In general, the fine-structure results in the broadening and shift of the spectra (overestimation of ion temperature and blue Doppler shift).

In high-temperature plasmas with ion temperatures in the keV range, the Doppler broadening is large enough to ignore the fine structure. However, the fine structure has a major impact on the evaluation of the plasma flow velocity from the Doppler shift.

Figure 3.22(b) shows the density dependence of the averaged wavelength of the charge-exchange line by the reaction with the thermal neutral and beam neutral. The averaged wavelength of the charge-exchange line by the reaction with thermal neutral, which contributes the cold component, is 3 pm larger than that with the beam neutral at low density, because of the difference in l distribution of the charge-exchange cross-section between thermal neutral and beam neutral charge exchange. Therefore, both the averaged wavelength of the thermal neutral charge and the beam neutral charge-exchange line start to decrease when the density exceeds 10^{19} m^{-3}. This is because the electron captured at each l level is redistributed by the collisions and the l distribution approaches the statistical weight of $(2l + 1)$ as the density is increased, a process called l mixing. At higher densities above 10^{20} m^{-3} the wavelength difference between these charge-exchange reactions disappears because of the l mixing. In the toroidal plasma density range of 10^{19}–10^{21} m^{-3}, the averaged wavelength is sensitive to the initial l distribution of the partial charge-exchange cross-section, which strongly depends on the atomic process model. Precise evaluation of the averaged wavelength of the charge-exchange line is difficult in experiment for two reasons: (i) the uncertainty in the atomic model used, and (ii) the stability of the absolute wavelength in the spectrometer. Therefore, bidirectional observation with two lines of sight contradictory to each other is required to measure the net Doppler shift and the precise plasma flow velocity.

3.3.4 Cross-section effect

Because the emission cross-section $Q(=\langle \sigma v \rangle_\lambda)$ has a strong energy dependence, the apparent Doppler shift due to the coupling between the gradient of the emission cross-section $(\partial Q/\partial v)$ and the gyro motion of ions in the plasma can be significant. As can be seen in figure 3.23, an ion traveling to the upstream of the beam with a velocity of V_{beam} has a larger relative velocity of $V_{beam} + |V_{ion}|$ than that of one traveling to the downstream of the beam $(V_{beam} - |V_{ion}|)$. The slope of the emission cross-section of the beam energy of 35 keV is positive. The emission from the traveling upstream ions becomes larger than the emission from the traveling [upstream] ions and causes an apparent Doppler shift parallel to the upstream of the neutral beam. Figure 3.23 shows the difference between the velocity distribution of the ion population and the charge-exchange emission of the carbon ions with an ion temperature, T_i, of 1 keV. Here, the positive V_{ion} is the velocity in the direction of the upstream of the beam. Because of the finite $(\partial Q/\partial v)$ as seen in figure 3.23(b), the charge-exchange emission in the positive side becomes larger than that in the negative side and causes the apparent Doppler shift by ~15 km s^{-1} parallel to the upstream of the neutral beam at the beam energy of 35 keV. The gradient of the emission cross-section $(\partial Q/\partial v)$ becomes negative at the higher beam energy, and the cross-section effect causes the apparent Doppler shift in the downstream direction of the neutral beam.

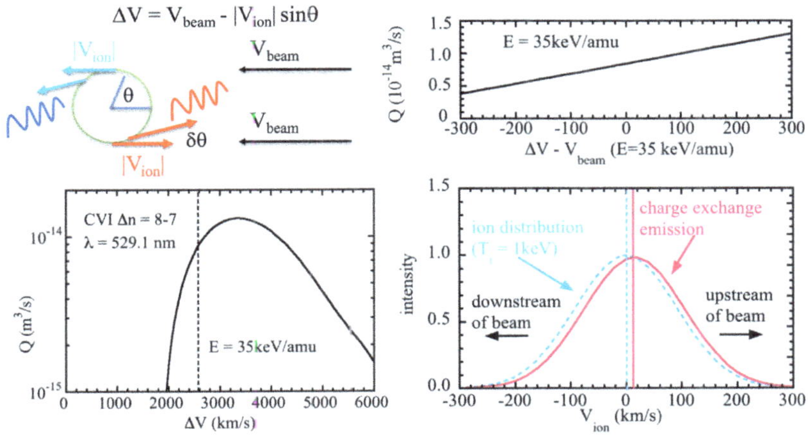

Figure 3.23. (a) Velocity dependence of emission charge-exchange cross-section $Q(=\langle\sigma v\rangle_\lambda)$, of (b) variation of Q at the beam velocity for $E = 35$ keV, (c) ion population of fully ionized carbon, and (d) charge-exchange recombination emission due to the reaction between fully ionized carbon and the hydrogen neutral beam.

This cross-section effect reaches its maximum for the ions traveling in the direction parallel to the beam ($\theta = \pi/2,\ 3\pi/2$) and its minimum for the ions traveling in the direction perpendicular to the beam ($\theta = 0,\ \pi$). Because of the finite lifetime of the transition from n to n', the direction of the emission is slightly tilted from the direction of the traveling ion by $\delta\theta$ due to the gyro motion, as seen in the figure. The lifetime is determined by the inverse of a coefficient at $\tau = 1/A$, and the tilted angle is determined by the lifetime τ and time of gyro motion (the inverse of the ion gyro frequency for the given magnetic field strength). Therefore, the apparent Doppler shift appears even if the emission is observed in the direction perpendicular to the beam velocity, which is called the secondary cross-section effect, while the apparent Doppler shift that appears parallel to the beam line is called the primary cross-section effect. In the measurement of flow velocity in the toroidal direction, a correction of the flow velocity due to the primary cross-section effect is necessary. In contrast, in the measurement of flow velocity in the poloidal direction, which is usually perpendicular to the beam line, a correction of the flow velocity due to the secondary cross-section effect may be important. The apparent Doppler shift due to the primary and secondary effects is roughly proportional to the magnitude of the ion temperature, as seen in figure 3.24. Here, the apparent velocity shift due to the primary effect parallel to the beam line is denoted by $V_{\|}$, while the apparent velocity shift due to the secondary effect perpendicular to the beam line is denoted by $V_{\perp}$. The magnetic field strength in this calculation is 3 T. In typical conditions in toroidal plasma, the secondary effect is smaller than the primary effect by an order of magnitude. However, the poloidal flow velocity is usually smaller than the toroidal flow velocity, and the secondary effect should also be taken into account.

In order to evaluate the real plasma flow velocity, the apparent flow velocity, $V_{\rm app}$, due to this effect should be subtracted. This subtraction is referred to in CXS as the cross-section effect correction. Correction of the measured plasma flow velocity and

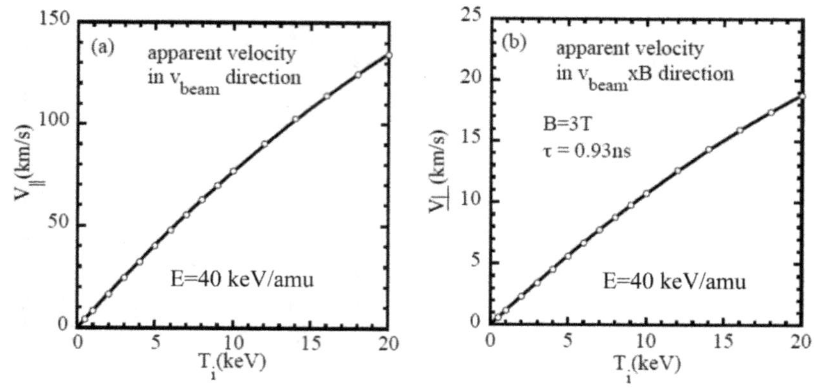

Figure 3.24. (a) Primary and (b) secondary charge-exchange cross-section effects at the beam velocity for the n = 8–7 transition of hydrogen-like carbon at E = 40 keV.

evaluation of the gradient of the emission cross-section ($\partial Q/\partial v$) is very important for precise measurement of the plasma flow velocity. Usually, the gradient of emission cross-section is evaluated from the Atomic Data and Analysis Structure (ADAS) database. Direct measurements of the gradient of emission cross-section have been also performed in experiments [30]. This apparent velocity, $V_{\rm app}$, is roughly proportional to the ion temperature. Therefore, $V_{\rm app}/T_{\rm i}$ can be determined in experiment based on the relation between the measured ion temperature and the measured flow velocity in the radial direction, where the real plasma flow velocity is zero.

In the low-energy region of the beam below 35 keV, the contribution of the excited neutral beam with the electrons in the $n = 2$ state starts to affect the ($\partial Q/\partial v$) values. The population of electrons at the $n = 2$ excited state of the neutral beam is typically <0.5%. However, the emission cross-section for the neutral beam in the excited state, $Q(n = 2)$, is much larger than that for the neutral beam in the ground state, $Q(n = 1)$, at low energy below $E>$ 20 keV. Therefore, even a 0.2–0.3% population of electrons in the $n = 2$ excited state can increase the emission cross-section because the $Q(n = 1)$ decreases sharply as the energy is decreased, as seen in figure 3.25(a). The effect of the $n = 2$ excited state on the normalized gradient of the emission cross-section, $(1/Q)(\partial Q/\partial v)$, becomes visible below 35 keV/amu and is significant at energies below 25 keV amu^{-1}. Figure 3.25(b) shows a comparison of the normalized gradient of the emission cross-section between that measured in experiment and that calculated with ADAS309. The experimental values show reasonable agreement with the ADAS calculation with the $n = 2$ excited states of the neutral beam of 0.2%.

3.3.5 Application to diagnostics

Charge-exchange spectroscopy has been widely used as a tool to measure ion temperature, plasma flow, and the density of impurities [33, 34]. Figure 3.26 shows radial profiles of ion temperature and flow velocity (toroidal rotation velocity) and carbon density measured with the C^{5+} line of 529.2 nm using multi-channel charge

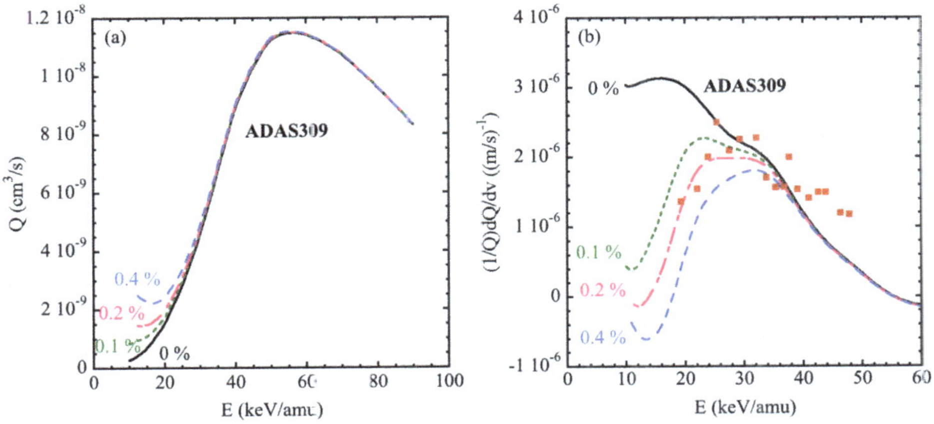

Figure 3.25. Energy dependence of (a) emission cross-section, and (b) normalized gradient of emission cross-section for the n = 8–7 transition of hydrogen-like carbon for various fractions of 0.1%, 0.2%, and 0.4% of the n = 2 neutral beam excited state. The experimental values are indicated with symbols in (b). Reprinted from [30], Copyright (2019), with permission from Elsevier.).

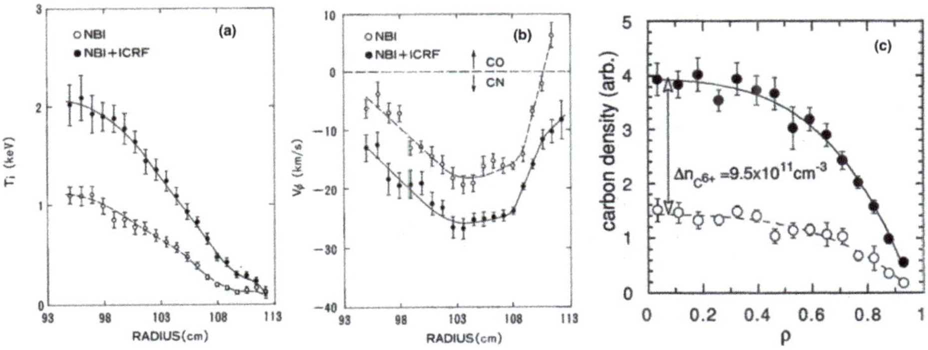

Figure 3.26. Radial profiles of (a) ion temperature and (b) flow velocity in the toroidal direction, and (c) the carbon density. Panels (a) and (b) reproduced from [31]. © International Atomic Energy Agency. Published by IOP Publishing. All rights reserved. Panel (c) reprinted from [32], Copyright (1992), with permission from Elsevier.

spectroscopy equipped with a CCD camera as a detector in the JIPP-TIIU [31]. In this system, 100 optical fibers (50 channels for the on-beam line and 50 channels for the off-beam line) are aligned at the entrance slit and the 2D CCD detector is introduced to measure the space- and wavelength-resolved intensity of the charge-exchange line emission. The major radius of the plasma center is 91 cm and that of plasma edge is 113 cm. The central ion temperature increases from 1 to 2 keV by introducing ICRH (marked ICRF in the figure) in addition to the NBI-heated plasma. The plasma flow velocity in the toroidal direction is anti-parallel to the plasma current (labeled as the CN direction) in the NBI-heated phase despite the fact that the NBI is injected to the direction parallel to the plasma current (labeled as the CO direction). After adding the ICRH (marked ICRF in the figure), the plasma flow is decelerated (or accelerated in the direction parallel to the plasma current). Figure 3.26(c) shows the carbon density for the hydrogen plasma (carbon exists in the plasma as an impurity) and the methane (CH_4) plasma [32].

The most significant advantage of CXS in the visible range is that optical fiber arrays can be used in the transmission of light from the plasma to the spectrometer. The measurement location in the plasma can be also easily scanned by the optics. The first and second derivative of the ion temperature in space was measured in the JT-60U by introducing modulation optics to the CXS [35]. The effective number of spatial channels is ~300. Because of the excellent spatial and temporal resolution of the measurement, CXS has been used to measure impurity density profiles, especially low-Z impurities where the impurity is fully ionized in the high-temperature plasma in a toroidal system. Impurity transport analysis using fully ionized impurities is free from uncertainties due to atomic processes such as ionization, recombination, and excitations. An example of impurity transport analysis is described in the section of 5.4.

It should be noted that the quantity measured with CXS can drive the radial electric field, which is a very important parameter to determine the heat and particle transport in toroidal plasma. The radial electric field (E_r) can be driven by the ion temperature, ion density, and plasma flow velocity perpendicular to the magnetic field as

$$E_r = \frac{\partial p_I}{eZ_I n_I \partial r} + V_\perp B, \tag{3.11}$$

$$V_\perp B = V_\phi B_\theta - V_\theta B_\phi, \tag{3.12}$$

where $V_\perp$, V_ϕ, and V_θ, are the plasma flow velocity perpendicular to the magnetic field, in the toroidal direction, and in the poloidal direction, respectively, and B, B_ϕ, and B_θ are the magnetic field strength and the toroidal and poloidal magnetic field, respectively. p_I, Z_I, and n_I are the pressure, electric charge, and the density of impurity I.

The validity of measurement of the radial electric field with CXS has been confirmed by comparison of the radial electric field obtained by the formula described above and that measured with a heavy ion beam probe (HIBP) in the CHS, which is equipped with both a HIBP system [37] and CXS [36]. The radial electric field profiles derived from the CXS and HIBP show good agreement within the error bars of the measurement, as seen in figure 3.27. The radial electric field is positive in the core region ($\rho < 0.5$), where the electron temperature is much higher than the ion temperature. This is known as electron root (electron loss is dominant) and the formation of a positive radial electric field is expected. Comparison of the two measurements shows that the radial electric field measured with CXS is smaller than that with HIBP by 10%, which exceeds the statistical scatter of the measurements. This discrepancy is considered as being due to the smoothing effect of CXS measurements in space due to its poor spatial resolution compared with HIBP.

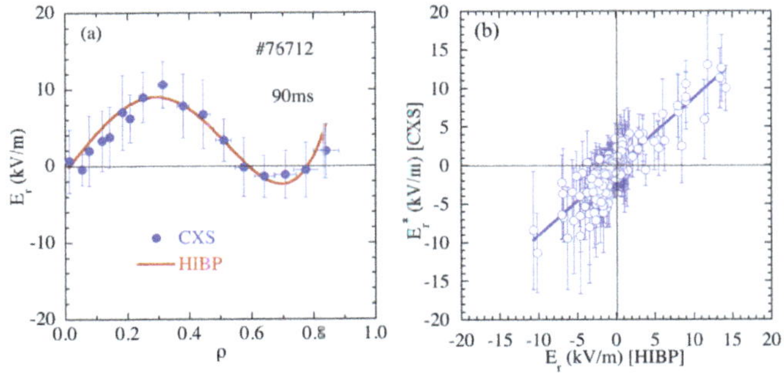

Figure 3.27. (a) Radial profile of the radial electric field measured with CXS and heavy ion beam probe (HIBP),and (b) comparison between these two measurements in the CHS. Reprinted from [36], with the permission of AIP Publishing.

Charge-exchange spectroscopy has made a great contribution to the finding of the radial electric field shear at the pedestal of the high-confinement mode (H-mode) in tokamak plasmas [38, 39]. The radial electric field near the plasma edge is weak in the low-confinement mode (L-mode) due to damping of the poloidal flow by large neoclassical viscosity. It has been noted that the radial electric field becomes more negative only near the plasma edge, which is called the E_r well, at the transition between the L-mode and H-mode phases. This E_r well is localized in a narrow region and has a large E_r shear (the gradient of E_r). This is one of the most significant findings in the study of transport in toroidal plasma, because the radial electric field shear is a key plasma parameter in suppressing turbulence and reducing transport. After helping to locate the radial electric field shear in the pedestal region of H-mode plasmas, CXS has been used widely in measurement of the radial electric field because of its good spatial resolution of up to a few millimeters and good time resolution of up to below 1 ms.

References

[1] Stratton B C, Bitter M, Hill K W, Hillis D L and Hogan J T 2017 Chapter 5: passive spectroscopic diagnostics for magnetically confined fusion plasmas *Fusion Sci. Technol.* **53** 431–86

[2] Morton A H and Srinivasacharya K G 1978 Use of optical filters for Tokamak plasma spectroscopy *J. Phys. E: Sci. Instr.* **11** 131–4

[3] Xu Z, Wu Z W, Gao W, Chen Y J, Wu C R and Zhang L *et al* 2016 Filterscope diagnostic system on the experimental advanced superconducting tokamak (EAST) *Rev. Sci. Instrum.* **87** 11D429

[4] Meyer O, Giacalone J C, Gouin A, Pascal J Y, Klepper C C and Fedorczak N *et al* 2018 Visible spectroscopy diagnostics for tungsten source assessment in the WEST tokamak: first measurements *Rev. Sci. Instrum.* **89** 10D105

[5] Meo F, Stansfield B L, Chartré M, de Villers P, Marchand R and Ratel G 1997 Spectroscopic imaging system for quantitative analysis of the divertor plasma of the Tokamak de Varennes *Rev. Sci. Instrum.* **68** 3426–35

[6] Harada T and Kita T 1980 Mechanically ruled aberration-corrected concave gratings *Appl. Opt.* **19** 3987–93

[7] Harada T, Kita T, Itou M, Taira H and Mikuni A 1986 Mechanically ruled diffraction gratings for synchrotron radiation *Nuclear Instruments and Methods in Physics Research* A **246** 272–7

[8] Zhou H, Morita S, Goto M, Narihara K and Yamada I 2010 Application of Z_{eff} profile analysis based on visible bremsstrahlung measurement to different density profiles in the LHD *Plasma Fusion Res.* **5** S1021–1

[9] Yoshinuma M, Ida K, Yokoyama M, Osakabe M and Nagaoka K 2010 Charge-exchange spectroscopy with pitch-controlled double-slit fiber bundle on LHD *Fusion Sci. Technol.* **58** 375–82

[10] Kobayashi M, Morita S and Goto M 2017 Space-resolved visible spectroscopy for two-dimensional measurement of hydrogen and impurity emission spectra and of plasma flow in the edge stochastic layer of LHD *Rev. Sci. Instrum.* **88** 033501

[11] Fonck R J, Ramsey A T and Yelle R V 1982 Multichannel grazing-incidence spectrometer for plasma impurity diagnosis: SPRED *Appl. Opt.* **21** 2115–23

[12] Lawson K D, Coffey I H, Zacks J and Stamp M FJET-EFDA Contributors 2009 An absolute sensitivity calibration of the JET VUV SPRED spectrometer *J. Instrum.* **4** P04013

[13] Dong C, Morita S, Goto M and Zhou H 2010 Space-resolved extreme ultraviolet spectrometer for impurity emission profile measurement in large helical device *Rev. Sci. Instrum.* **81** 033107

[14] Gray D S, Luckhardt S C, Chousal L, Gunner G, Kellman A G and Whyte D G 2004 Time resolved radiated power during tokamak disruptions and spectral averaging of AXUV photodiode response in DIII-D *Rev. Sci. Instrum.* **75** 376–81

[15] Ross P A 1928 A new method of spectroscopy for faint X-radiations *J. Opt. Soc. Am.* **16** 433–7

[16] Stutman D, Finkenthal M, Delgado-Aparicio L, Tritz K, Tamura N and Kalinina D *et al* 2005 High throughput ultrasoft X-ray polychromator for embedded impurity pellet injection studies *Rev. Sci. Instrum.* **76** 013508

[17] Ming T F, Ohdachi S, Sakakibara S and Suzuki Y 2012 High speed vacuum ultraviolet telescope system for edge fluctuation measurement in the large helical device *Rev. Sci. Instrum.* **83** 10E513

[18] Brandt C, Schilling J, Thomsen H, Broszat T, Laube R and Schröder T *et al* 2020 Soft X-ray tomography measurements in the Wendelstein 7-X stellarator *Plasma Phys. Control. Fusion* **62** 035010

[19] Hollmann E M, Chousal L, Fisher R K, Hernandez R, Jackson G L and Lanctot M J *et al* 2011 Soft X-ray array system with variable filters for the DIII-D tokamak *Rev. Sci. Instrum.* **82** 113507

[20] Ingesson L C, Alper B, Peterson B J and Vallet J C 2017 Chapter 7: tomography diagnostics: bolometry and soft-X-ray detection *Fusion Sci. Technol.* **53** 528–76

[21] Lerche E, Goniche M, Jacquet P, Van Eester D, Bobkov V and Colas L *et al* 2016 Optimization of ICRH for core impurity control in JET-ILW *Nucl. Fusion* **56** 036022

[22] Hill K W, Bitter M, Diesso M, Dudek L, von Goeler S and Hayes S *et al* 1985 Tokamak fusion test reactor prototype X-ray pulse-height analyzer diagnostic *Rev. Sci. Instrum.* **56** 840–2

[23] Hill K W, von Goeler S, Bitter M, Campbell L, Cowan R D and Fraenkel B *et al* 1979 Determination of Fe charge-state distributions in the Princeton large torus by Bragg crystal X-ray spectroscopy *Phys. Rev.* A **19** 1770–9

[24] Weinheimer J, Ahmad I, Herzog O, Kunze H J, Bertschinger G and Biel W *et al* 2001 High-resolution X-ray crystal spectrometer/polarimeter at torus experiment for technology oriented research-94 *Rev. Sci. Instrum.* **72** 2566–74

[25] Ince-Cushman A, Rice J E, Bitter M, Reinke M L, Hill K W and Gu M F *et al* 2008 Spatially resolved high resolution X-ray spectroscopy for magnetically confined fusion plasmas (invited) *Rev. Sci. Instrum.* **79** 10E302

[26] Tokar M Z, Rapp J, Bertschinger G, Konen L, Koslowski H R and Kramer-Flacken A *et al* 1997 Nature of high-Z impurity accumulation in tokamaks *Nucl. Fusion* **37** 1691–708

[27] Lipschultz B, Labombard B, Marmar E S, Pickrell M M, Terry J L and Watterson R *et al* 2011 Marfe: an edge plasma phenomenon *Nucl. Fusion* **24** 977–88

[28] Fonck R J, Darrow D S and Jaehnig K P 1984 Determination of plasma-ion velocity distribution via charge-exchange recombination spectroscopy *Phys. Rev.* A **29** 3288–309

[29] Isler R C 1994 An overview of charge-exchange spectroscopy as a plasma diagnostic *Plasma Phys. Control. Fusion* **36** 171–208

[30] Chen J, Ida K, Yoshinuma M, Murakami I, Kobayashi T and Ye M Y *et al* 2019 Effect of energy dependent cross-section on flow velocity measurements with charge exchange spectroscopy in magnetized plasma *Phys. Lett.* A **383** 1293–9

[31] Ida K, Kawahata K, Toi K, Watari T, Kaneko O and Ogawa Y *et al* 1991 Observation of toroidal plasma rotation driven by the electric field induced by loss of ions *Nucl. Fusion* **31** 943–7

[32] Ida K and Kato T 1992 Line-emission cross sections for the charge-exchange reaction between fully stripped carbon and atomic hydrogen in tokamak plasma *Phys. Lett.* A **166** 35–9

[33] Groebner R J, Brooks N H, Burrell K H and Rottler L 1983 Measurements of plasma ion temperature and rotation velocity using the He II 4686 Å line produced by charge transfer *Appl. Phys. Lett.* **43** 920–2

[34] Ida K and Hidekuma S 1989 Space- and time-resolved measurements of ion temperature with the CVI 5292 Å charge-exchange recombination line after subtracting background radiation *Rev. Sci. Instrum.* **60** 867–71

[35] Ida K, Sakamoto Y, Yoshinuma M, Inagaki S, Kobuchi T and Matsunaga G *et al* 2008 Measurement of derivative of ion temperature using high spatial resolution charge exchange spectroscopy with space modulation optics *Rev. Sci. Instrum.* **79** 053506

[36] Ida K, Fujisawa A, Iguchi H, Yoshimura Y, Minami T and Okamura S *et al* 2001 Experimental test of the radial force balance equation in the compact helical system *Phys. Plasmas* **8** 1–4

[37] Fujisawa A, Iguchi H, Idei H, Kubo S, Matsuoka K and Okamura S *et al* 1998 Discovery of electric pulsation in a toroidal helical plasma *Phys. Rev. Lett.* **81** 2256–9

[38] Groebner R J, Burrell K H and Seraydarian R P 1990 Role of edge electric field and poloidal rotation in the L-H transition *Phys. Rev. Lett.* **64** 3015–8

[39] Ida K, Hidekuma S, Miura Y, Fujita T, Mori M and Hoshino K *et al* 1990 Edge electric-field profiles of H-mode plasmas in the JFT-2M tokamak *Phys. Rev. Lett.* **65** 1364–7

Chapter 4

Approaches to the study of impurity transport

Two types of experimental approaches to the study of impurity transport are discussed in this chapter. One approach is to analyze the impurity emission and density profiles during the recovery phase after the impurity redistribution event; the other approach is to analyze the impurity emission and density profiles after an external perturbation, such as impurity injection. The absolute values of the impurity transport coefficients (e.g., diffusion coefficient and convection velocity) are derived from the time-varying radial profile of the impurity density; however, measurements of the steady-state radial profile of the impurity density can only derive the ratio of convection velocity and diffusion coefficient.

4.1 Approaches with an intrinsic impurity redistribution

As aforementioned, in order to derive the transport coefficients (diffusion coefficient and convection velocity), the time-varying radial profile of the impurity density for a quasi steady-state background temperature and density cannot be derived from the steady-state radial profile of the impurity density. This is because the steady-state radial profile of the impurity density derives only the ratio of convection velocity and diffusion coefficient, and not the absolute values of these coefficients. Therefore, impurity injection, such as in laser blow-off experiments, has been widely performed in order to derive the diffusion coefficient and convection velocity. Another approach to impurity transport study is to evaluate the diffusion coefficient and convection velocity in the plasma where the intrinsic impurity redistribution takes place. The timescale of this impurity redistribution should be much shorter than the impurity transport timescale. Collapse events due to magnetohydrodynamics (MHD) instabilities cause the redistribution of the impurity profile. This approach has the advantage that an external perturbation, such as an impurity injection, is not required. However, the plasma condition and region for the impurity transport study are limited. The background density and temperature profiles also change in time during the recovery of the impurity profile before the events. The other approach is

doi:10.1088/978-0-7503-1451-0ch4

to take advantage of a rapid change of transport state, a so-called transition event. There are several states of transport with different bulk plasma particle diffusivity, heat diffusivity, and momentum diffusivity. One such plasma state with larger diffusivity is the low-confinement mode (L-mode), while some plasma states with smaller diffusivity are the high-confinement mode (H-mode) and an internal transport barrier (ITB). Impurity transport after a transition (typically the transition from an L-mode to a H-mode state or an L-mode to an ITB state) can be determined by this approach. Before the transition, only the ratio of convection velocity and diffusion coefficient can be determined from the steady-state profile.

4.1.1 Impurity transport analysis using sawtooth oscillation

Sawtooth oscillation, which is characterized by a slow rise of electron temperature followed by a rapid drop in temperature, is commonly observed in tokamak plasmas. The rapid drop of electron temperature is a consequence of the reconnection of the magnetic field of the well-known $m = 1$ and $n = 1$ tearing instability at the $q = 1$ rational surface, which is referred to as the magnetic island. Here, m and n are the toroidal and poloidal mode numbers. Figure 4.1 shows the time evolution of the central soft X-ray (SXR) intensity and a tomographically reconstructed two-dimensional (2D) X-ray emission profile [1]. The timing of the tomographically reconstructed 2D X-ray emission profile (at times E, F, G, H) is indicated in the expanded view. At time slice E, the 2D X-ray emission profile is peaked with a significant $m = 1$ component, which indicates the growth of the magnetic island. After the sawtooth crash (time slice H), a flattening of the 2D X-ray emission profile (or a slightly hollow profile) is observed and the $m = 1$ component also disappears. The timescale of the sawtooth crash (rapid drop of electron temperature) is on the order of 10^{-4} s, while the sawtooth period (slow rise of electron temperature) is on the order of 10^{-1} s, which is comparable to the timescale of impurity transport. The sawtooth crash causes a flattening of the impurity density profile as well as the temperature profile in a timescale much shorter than the impurity transport, and the sawtooth period is long enough for the impurity profile to recover to the profile determined by the transport process. The region of flattening of the density and temperature is observed inside the so-called mixing radius, which is usually 1.4 times the radius of the $q = 1$ rational surface. The mixing radius depends on the q profile, and is typically half the plasma minor radius. Therefore, sawtooth oscillation is the ideal perturbation for impurity transport study in the core region.

Both the diffusion coefficient and convection velocity can be derived from the time evolution of the radial profiles of the impurity density, after the flattening of the impurity density, by introducing a sawtooth crash model into the impurity transport code [2–4]. Figure 4.2 shows (a) the time evolution of the central Z_{eff} and nickel (Ni) and iron (Fe) densities, and (b) the radial profile of Z_{eff} in a discharge with a sawtooth crash in PBX tokamak plasma. The central Z_{eff} value shows the sharp drop in each sawtooth crash of 600 ms and 650 ms. Flattening of the Z_{eff} profile is observed inside the sawtooth inversion radius of 10 cm. After the sawtooth, the central Z_{eff} value starts to increase due to the inward convection of the impurity

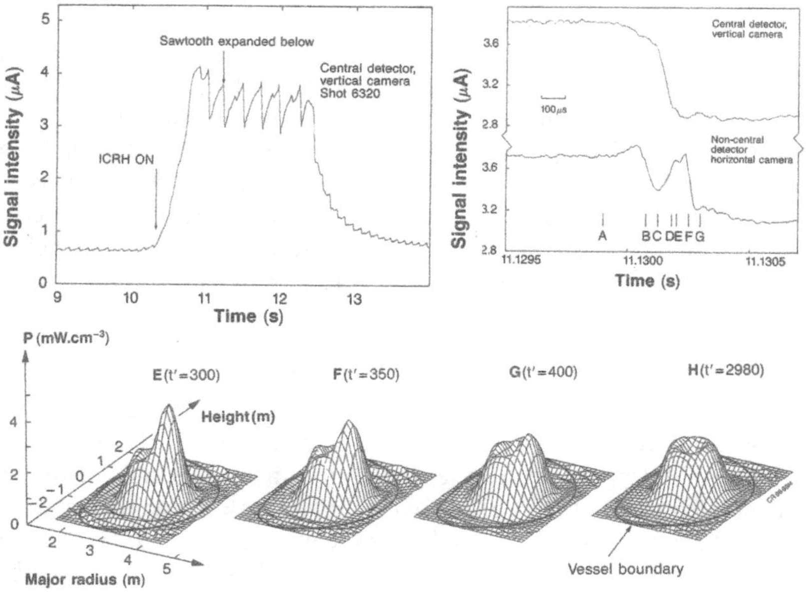

Figure 4.1. Time evolution of central soft X-ray intensity and tomographically reconstructed 2D X-ray emission profile. Reprinted with permission from [1], Copyright (1986) by the American Physical Society.

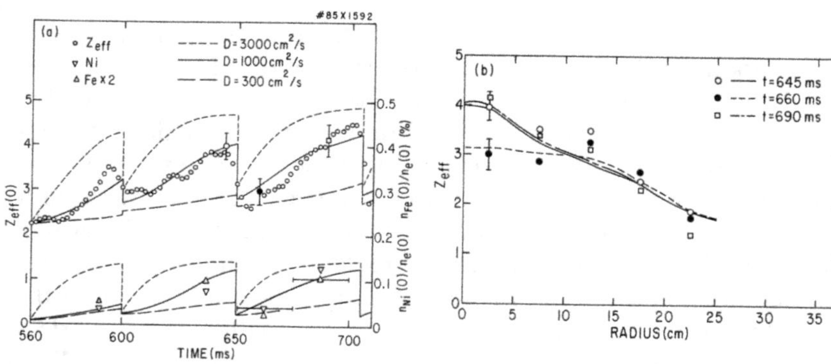

Figure 4.2. (a) Time evolution of central Z_{eff} and nickel (Ni) and iron (Fe) densities, and (b) radial profile of Z_{eff} just 5 ms before (t = 645 ms) and 10 ms and 40 ms after (t = 660 and 690 ms) the sawtooth crash. Simulation results with different coefficient values of 300, 1000, and 3000 cm^2 s^{-1} are also plotted. Reprinted with permission from [2], Copyright (1987) by the American Physical Society.

transport. Simulation results with three diffusion coefficients of 300, 1000, and 3000 cm^2 s^{-1} with a constant ratio of convection to diffusion, that is, peaking factor, c_v, are plotted. The peaking factor c_v is determined to reproduce the radial profile of the peaked Z_{eff} profile and radial profile just before the sawtooth crash. The peaking factor, defined as $c_v = -Va^2/(2Dr)$, is 4 for low-Z impurities (mainly carbon and oxygen) and 20 for high-Z impurities (mainly Ni and Fe). Because the minor radius a = 0.36 m and sawtooth radius r = 0.14 m, the convection velocity is 0.9 m s^{-1} for

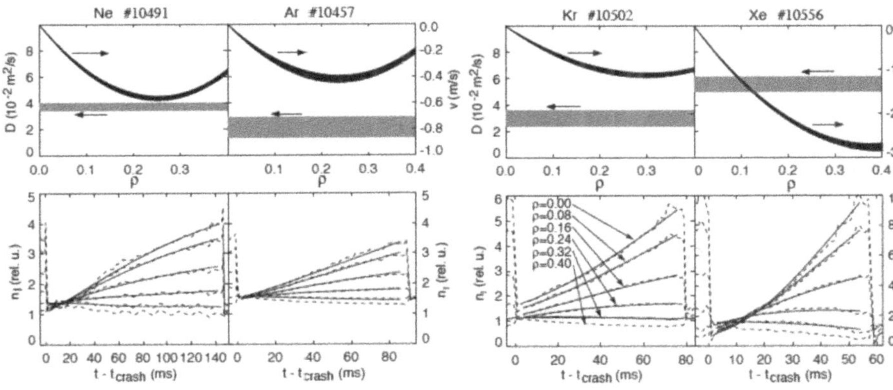

Figure 4.3. Radial profiles of diffusion coefficient (D) and convection velocity (V) and time evolution of impurity density at a normalized minor radius of $\rho = 0.0$, 0.08, 0.16, 0.24, 0.32, and 0.40 during the recovery phase after the sawtooth crash for neon (Ne), argon (Ar), krypton (Kr), and xenon (Xe) in the ASDEX Upgrade. The dashed lines are measurements and the solid lines are simulation results using the transport coefficients D and V. Reproduced from [6].

low-Z impurities and 4.4 m s^{-1} for high-Z impurities. Simulation results for a diffusion coefficient of 0.1 m^2 s^{-1} show good agreement with measurements. These were the first experimental results to show the Z dependence of convection velocity, as predicted by neoclassical theory. Since this experiment, the role of sawtooth oscillation in helping to avoid impurity accumulation has been recognized. [5] The unique feature of impurity transport analysis using a sawtooth crash is that the diffusion coefficient and convection velocity inside a $q = 1$ rational surface can be simultaneously measured for both low-Z and high-Z impurities.

Perturbation using a sawtooth has also been used to evaluate the diffusion coefficient and convection velocity in the ASDEX Upgrade tokamak. Figure 4.3 shows the radial profile of the diffusion coefficient (D) and convection velocity (V) and time evolution of the impurity density at a normalized minor radius of $\rho = 0.0$, 0.08, 0.16, 0.24, 0.32, and 0.40 during the recovery phase after the sawtooth crash for neon (Ne), argon (Ar), krypton (Kr), and xenon (Xe). The dashed lines are measurements and solid lines are simulation results using the transport coefficients D and V above. The diffusion coefficient and convection velocity are selected to reproduce the time evolution of the impurity density after the sawtooth crash for all radii. The diffusion coefficient is in the range 0.02–0.06 m^2 s^{-1}. The convection velocity for medium-Z impurities (Ne, Ar) is 0.2–0.4 m s^{-1}, while the convection velocity for high-Z impurities (Kr, Xe) is 1.0–3.0 m s^{-1}. With increasing Z number, the transport becomes strongly convective with inward-directed drift velocity. This Z dependence of impurity transport is consistent with that observed previously in the PBX. Experiments in the PBX and JET demonstrated that the diffusion and convection velocity in the core region (i.e., inside the $q = 1$ rational surface) can be evaluated from the time evolution of the impurity density during the recovery phase after the sawtooth crash. This is in contrast to impurity transport analysis using an impurity injection (described in the next section), which is useful in evaluating the diffusion coefficient and convection velocity outside the $q = 1$ rational surface (i.e., in the outer half of the plasma minor radius).

4.1.2 Impurity transport analysis using edge-localized modes

Edge-localized modes (ELMs) are periodic rapid energy and particle loss events commonly observed in the H-mode pedestal. These periodic losses are restricted to the region near the plasma boundary and cause periodic spikes of D_α intensity. There are three types of ELM depending on the parameter regime. Type I ELMs are characterized by large, discrete spikes of D_α intensity due to the sudden increase of recycling associated with the abrupt loss of energy and particles, and appears in the parameter region close to the ideal ballooning limit. In contrast, Type II and Type III ELMs are characterized by small and more frequent spikes. Discharges with Type I ELMs are commonly used in impurity transport studies. The time interval of the periodic losses in Type I ELMs is 10^{-2}–10^{-1} s, while the timescale of losses is 10^{-3} s, which is the ideal timescale for perturbations in the study of impurity transport. As can be seen in figure 4.4, a simultaneous decrease of electron density inside the last closed-flux surface (LCFS) and increase of electron density outside the LCFS are observed in a Type I ELM event. The density pivot point where the electron density is unchanged is located at the LCFS. The change in the density gradient across the LCFS occurs within a half millisecond and is restricted to near the plasma boundary of 0.8–1.1 of the normalized minor radius.

ELM events cause a redistribution of the impurity density profile as well as the electron density. Figure 4.5 shows radial profiles of carbon (C) density near the LCFS from 5 ms before to 5 ms after the ELM event, in addition to radial profiles of the diffusion coefficient and the ratio of convection velocity to diffusion coefficient for the modeling to fit the experimental data [8]. Before the ELM, a very sharp C density gradient is observed. At the ELM event, this sharp gradient disappears, but quickly recovers within a few milliseconds. In order to reproduce this sharp C density gradient at the LCFS with a diffusion/convection model, a large inward convection velocity (negative V/D) localized at the LCFS is required. The magnitude of the V/D ratio at the LCFS is larger than that observed in an ITB, as described in chapter 7.3, by one order of magnitude. This is because the mechanism causing the large impurity gradient at the LCFS is different from that in an ITB region, where the suppression of turbulence plays an important role. The negative convection velocity becomes more significant for high-Z impurities. At the time of the ELM event, the effective diffusion coefficient to reproduce the experimental observation exceeds 10 $m^2\ s^{-1}$, which is almost two orders of magnitude larger than that evaluated from the impurity density during the recovery phase between ELM intervals. This large effective diffusion coefficient is not due to the collision process of impurity ions but to the radial displacement of impurity ions driven by $E_\theta \times B$, associated with the MHD collapse event. The diffusion coefficient inside the pedestal top ($\rho < 0.95$) is $\sim 1\ m^2\ s^{-1}$, which is a typical magnitude predicted by the turbulence in so-called anomalous transport.

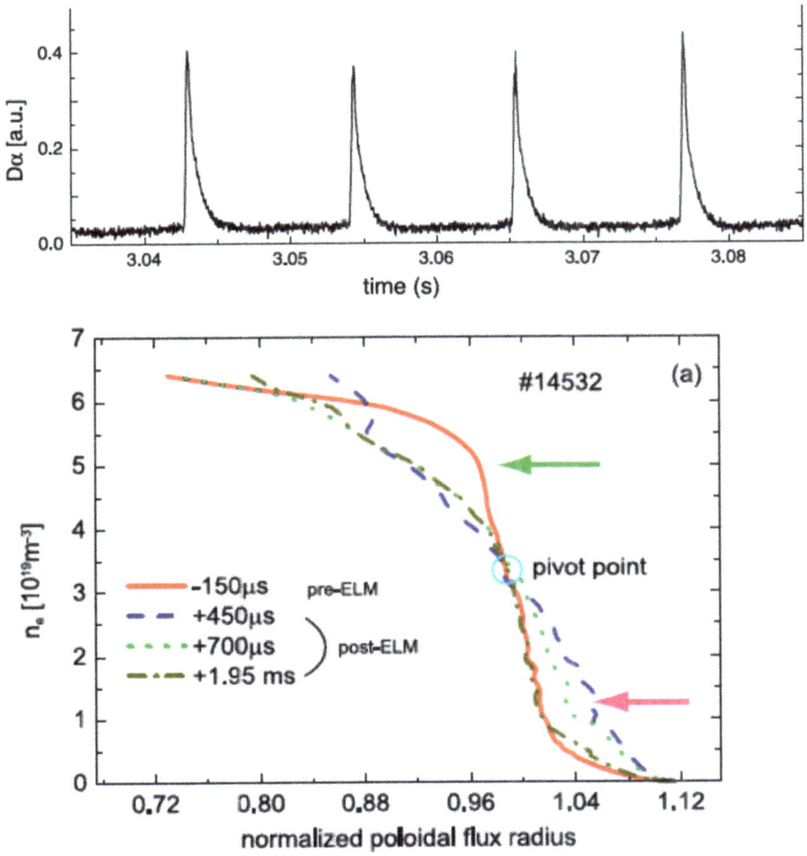

Figure 4.4. Time evolution of D_α intensity and radial profiles of electron density before and after the ELM event. Reproduced from [7].

The disadvantage of impurity transport study using a sawtooth oscillation is that evaluation of the transport coefficients is restricted to the core region (typically only half of the plasma minor radius), because the flattening of the impurity density due to the sawtooth crash only takes place inside the $q = 1$ rational surface. In contrast, ELMs are MHD instabilities localized near the plasma edge that cause a significant change in the electron density and impurity density profiles. Thus, ELM events are characterized by a sudden loss of plasma at the edge region, triggered by the MHD instability, and are usually observed by a sharp spike of Hα emission. As seen in figure 4.6, the Ni XIV and Ni XVIII emission lines with Ni ions in a low charge state show sudden drops in each ELM event, while the Ni XXV emission of a higher charge state shows a gradual increase after each ELM event. These clearly observed behaviors indicate the strong inward convection that is expected in neoclassical impurity transport.

The radial profiles of the impurity transport coefficients (diffusion coefficient and convection velocity) evaluated from the time evolution of the Ni XIV, Ni XVIII, and

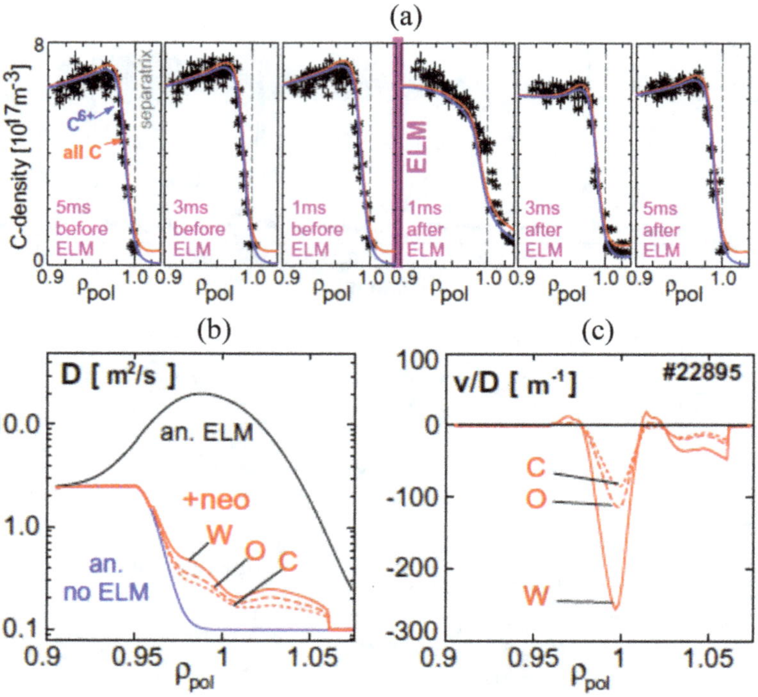

Figure 4.5. (a) Radial profiles of carbon (C) density near the LCFS from 5 ms before to 5 ms after the ELMs event, in addition to radial profiles of (b) anomalous diffusion coefficient, and (c) the ratio of convection velocity to diffusion coefficient for the modeling to fit the experimental data. Reprinted from [8], Copyright (2011), with permission from Elsevier.

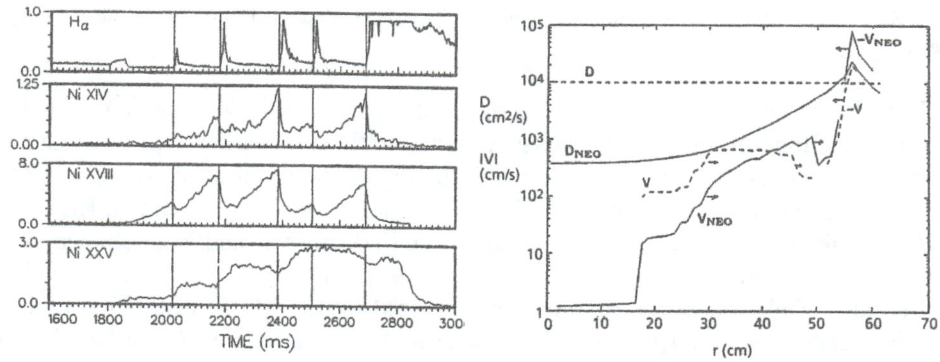

Figure 4.6. Time evolution of H_α intensity and intrinsic nickel intensities of Ni XIV, Ni XVIII, Ni XXV lines in a H-mode discharge with ELM (left), and radial profiles of diffusion coefficient and convection velocity (right) evaluated during the recovery phase (dashed lines) and with neoclassical values (solid lines). Reproduced from [9]. © International Atomic Energy Agency. Published by IOP Publishing. All rights reserved.

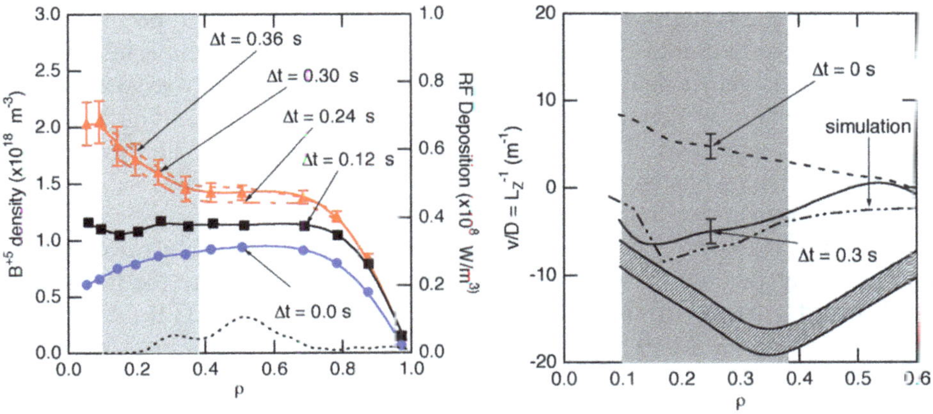

Figure 4.7. Radial profiles of boron density B^{+5} during the formation of an ITB (left) and radial profiles of the inverse scale length for the impurity (right). Here, δt is the time elapsed following ITB formation. Reproduced from [11].

Ni XXV emission lines are compared with that predicted by neoclassical theory. Here, the diffusion coefficient is assumed to be constant in space. The large increase of convection velocity near the plasma edge causes a severe problem of impurity accumulation in H-mode plasmas between ELM events. Because impurities tend to accumulate in H-mode plasmas due to the large inward convection at the pedestal region, the exhaust of impurities in H-mode discharges is a crucial issue, especially in metal-wall devices. When the impurity levels at the plasma edge are reduced, it contributes to a reduction of inward convection due to collisions between impurities. Therefore, the reduction of edge impurity concentration with small ELMs contributes to both lower impurity density at the plasma edge and lower peaking of the impurity density profile.

4.1.3 Impurity transport analysis using the transition from a low-confinement mode to an internal transport barrier

The other approach to the study of impurity transport is by analysing the impurity density profiles after the formation of an ITB. In this approach, there are two evaluation techniques. One is to derive the transport coefficients (diffusion coefficient and/or convection velocity) by comparison of the time evolution of the measured impurity transport with that simulated using an impurity transport code. The other is to derive the transport coefficient directly from the flux gradient relation after the formation of the ITB [10]. Direct evaluation of the transport coefficient requires precise measurements of impurity density profiles with high spatial resolution, because the gradient of the impurity density is necessary for these evaluations. In contrast, evaluation using simulation code does not require precise impurity density measurements.

Figure 4.7 shows radial profiles of boron density B^{+5} during the formation of an ITB and radial profiles of the inverse scale length for the impurity [11]. The boron

(B) density profile is slightly hollow before the ITB formation, and becomes peaked 0.3 s after the formation of ITB. The ITB-induced impurity gradient is interior to the ion cyclotron resonance heating deposition (dotted black line, right axis) and interior to the ITB region as defined using the electron density profile. The ITB region ($0.3 < \rho < 0.4$) is shaded. The change of B profile from slightly hollow to peaked indicates the change of sign of the convective velocity. Here, the inverse scale length for the impurity is treated as an empirical V/D, wherein a negative value represents pinch-like inward convection, and a positive value represents outward convection. These data suggest that the convection velocity is outward before the formation of the ITB and switches to inward after the formation of the ITB. A numerical simulation at $\delta t = 0.3$ s is shown as a dashed–dotted curve and a neoclassical prediction, $V_{\mathrm{neo}}/D_{\mathrm{neo}}$, is shown as a hatched band. The inward convection observed is much smaller than that predicted by neoclassical theory.

Figure 4.8 shows an example comparison of measurements and simulations of ITB plasma during the impurity accumulation phase ($t = 6.2$–6.8 s) after the formation of the ITB in the JET tokamak [12]. In this analysis, the convection velocity is assumed as a neoclassical convection velocity ($V_z^{\mathrm{conv}} = V_{\mathrm{neo}}$) and the diffusion coefficient is the sum of the neoclassical diffusion coefficient (D_{neo}) and anomalous diffusion coefficient (D_{an}), as $D_z = D_{\mathrm{neo}} + D_{\mathrm{an}}$. In this experiment, the neoclassical diffusion and convection velocity (D_{neo} and V_{neo}) are calculated from neoclassical code using the measured density and temperature profiles, while the anomalous diffusion coefficient (D_{an}) is chosen to reproduce the measured impurity profile. As can be seen in figures 4.8(c) and (d), the neoclassical diffusion coefficients for Ne and Ni are smaller than the anomalous diffusion coefficient, except for the ITB region ($r = 0.2$–0.6 m), while a large negative (inward) convection velocity appears inside the ITB region. The anomalous diffusion coefficient is assumed to have no Z dependence, the same as in previous research [2]. The radial profile of the anomalous diffusion coefficient is selected to match the time evolution ($t = 6.2$, 6.6, 6.8 s) of the radial profile both for Ne and Ni, as seen in figures 4.8(e) and (f). Both of the simulated Ne and Ni density profiles show reasonable agreement with the measured impurity density profiles. The central diffusion coefficient ($r < 0.2$ m) is set to $D_{\mathrm{an}} = 1\ \mathrm{m^2\ s^{-1}}$. The diffusion coefficient in the ITB region ($r = 0.2$–0.6 m) drops significantly to very low values of $D_{\mathrm{an}} = 0.02\ \mathrm{m^2\ s^{-1}}$, and the low-diffusivity region expands in time ($t = 6.2$–6.8 s) as the ITB develops. The low diffusion coefficient inside the ITB region observed in the JET is consistent with the low diffusion coefficient evaluated from the charge state distribution of titanium (Ti) ions inside the electron ITB in the CHS, described in chapter 2.2.

4.2 Approaches with a non-intrinsic impurity injection

This section discusses the experimental approaches for the study of impurity transport using the injection of a non-intrinsic impurity (i.e., an external perturbation). In these approaches, impurities from outside the vacuum vessel are injected to artificially invoke the time-varying radial profile of the impurity ions, which is essential in estimating the transport coefficients (diffusivity coefficient and

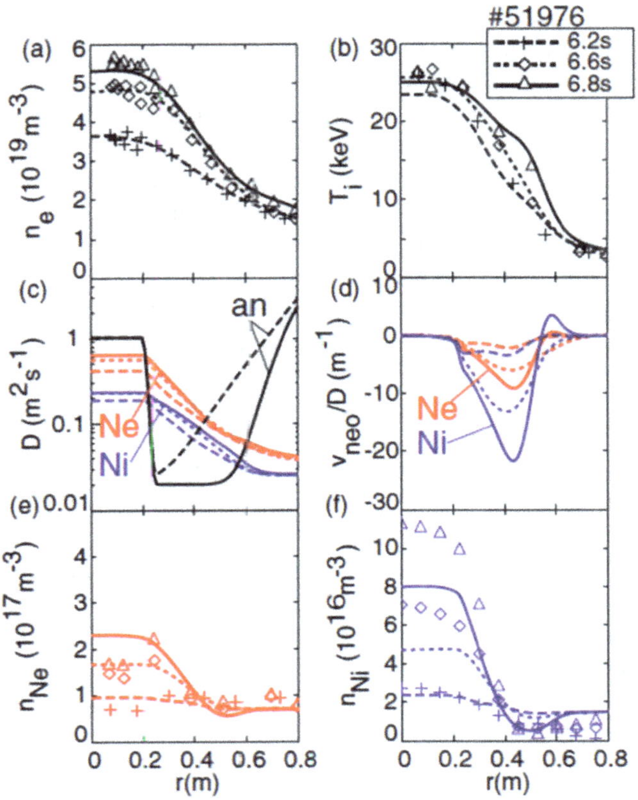

Figure 4.8. Three radial profiles of (a) electron density, (b) ion temperature, (c) neoclassical and anomalous diffusion coefficient, D_{neo}, D_{an}, (d) neoclassical convection velocity normalized total diffusion coefficient $V_{neo}/(D_{neo} + D_{an})$, (e) neon (Ne) density, and (f) nickel (Ni) density from the impurity transport simulation in ITB plasma in JET. Overlaid symbols give experimentally measured profiles. Reproduced from [12].

convection velocity) individually. One of the advantages of such approaches is that the study of impurity transport can be performed at almost any time during the discharge. With this advantage, it is possible to expand the research of impurity transport in magnetically confined fusion (MCF) plasmas. The excessive radiation from impurity ions occasionally leads to the termination of MCF plasmas, mainly caused by the accumulation of impurities toward the plasma center. It is thus crucially important to understand the physical mechanisms underlying this impurity accumulation in MCF plasmas. To this end, it is critical to study the impurity transport just before or during the impurity accumulation. Another advantage of an external impurity injection into a MCF plasma is the utilization of non-intrinsic impurities. As already discussed in the previous section, in impurity transport studies utilizing the impurity redistribution in the plasmas, the impurities to be investigated need to be spread throughout the plasma before the rapid change (i.e., transition) of the plasma structure. From this point of view, only the intrinsic impurities can serve as the target of such studies utilizing the redistribution of impurity ions in plasmas. Usually, the intrinsic impurities, e.g., carbon, and some metallic impurities, as typified by iron, originate from the plasma-facing components (PFCs) in the vacuum vessel. Thus, the kinds of intrinsic impurities that can be investigated are limited. In contrast, the external injection of non-intrinsic impurities can dramatically increase the scope of the impurities to be investigated. The study of an expanded number of impurities has the merit that the Z dependence of impurity transport can be easily studied. It is essential to study this Z dependence of impurity transport in order to extract experimentally the physical parameters governing impurity transport in MCF plasmas.

The impurities injected into MCF plasmas are either quickly or slowly expelled from the plasmas, except in the presence of impurity accumulation. However, some impurities can re-enter the plasma again after their initial expulsion. This phenomenon is known as recycling. Therefore, since the impurity ion density profile in a MCF plasma is affected not only by the impurity injection rate but also the wall-recycling rate, any impurity transport analysis after the injection of the non-intrinsic impurity should take into account such a wall-recycling rate, which otherwise could increase uncertainty in the analysis. On the other hand, the injection of non-recycling non-intrinsic impurities can simplify the impurity transport study in MCF plasmas. In particular, the impurity particle confinement time can be derived from the exponential time decay of line emissions from highly ionized impurities, which can exist in the core region of MCF plasmas after their peaks. The impurity particle confinement time analysis requires a stationary plasma condition during the decay phase of the impurity line emissions.

Three typical external impurity injection schemes have been widely utilized in current-generation MCF experiments: gas puffs, laser blow-off (LBO), and impurity pellets. Each scheme is discussed in more detail below.

4.2.1 Impurity transport analysis with a gas puff

A gas-puff scheme has been widely utilized in impurity transport studies in MCF experiments. In this scheme, a noble gas such as helium (He), Ne, or Ar is typically puffed into the MCF plasma.

However, since noble gas species are chemically inert, the noble gases can quickly re-enter the plasmas again after expulsion. Figure 4.9(a) shows the temporal evolution of Ar atoms in Alcator C-Mod plasma after a short Ar gas puff [13]. Measurement of the Ar XVII emission line with a crystal spectrometer estimates the number of Ar atoms. As can be immediately recognized, the number of Ar atoms injected into the plasma increased even after the Ar gas puff was ended. This relatively slow increment of Ar atoms in the plasma is mainly attributed to recycling, and the number of Ar atoms in the plasma is proportional to the integrated number of Ar atoms puffed into the vacuum vessel of the Alcator C-Mod. Therefore, the study of impurity transport with the injection of recycling gases requires a very short and intense gas injection. While nitrogen (N) is also chemically inert in the gaseous phase, it can form highly chemically reactive radicals in plasma. Hence, nitrogen behaves like a non-recycling impurity. As shown in figures 4.9(b) and (c), the

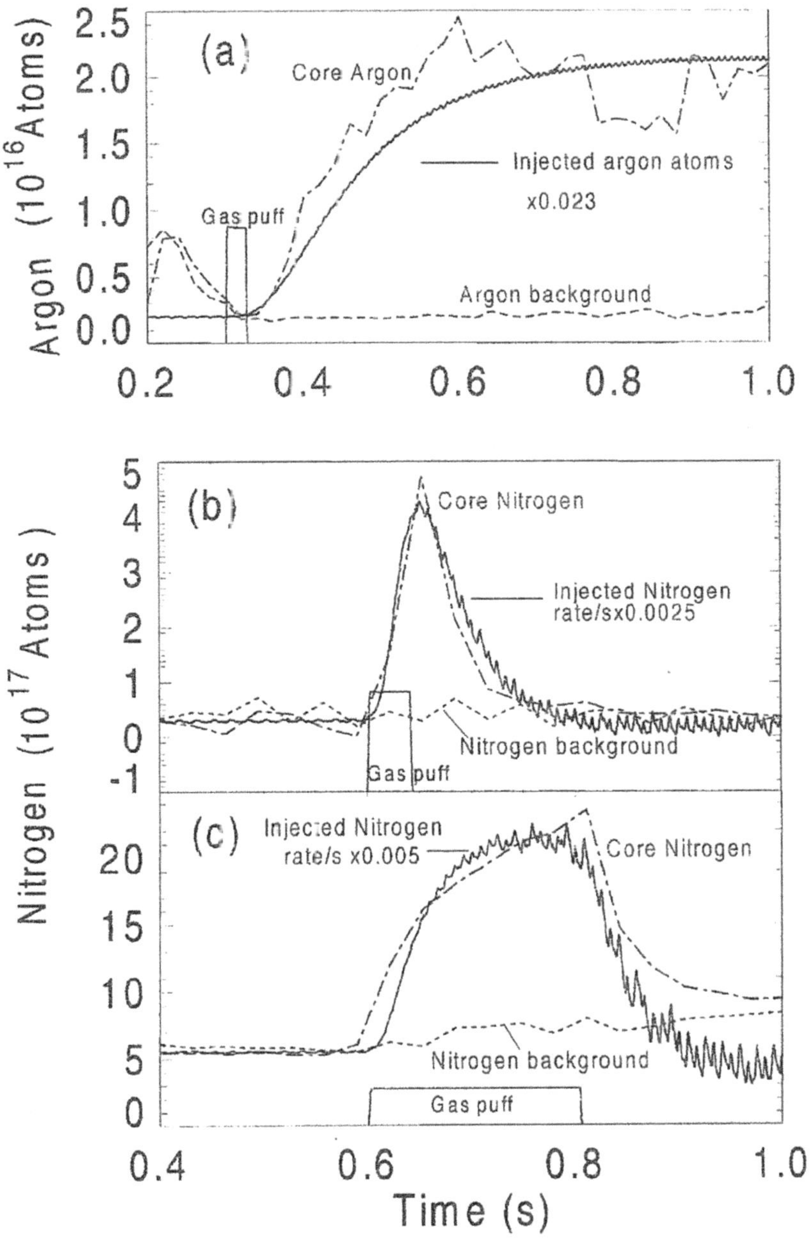

Figure 4.9. Temporal behaviors of (a) argon (Ar) atoms, and (b) nitrogen (N) atoms injected into the Alcator C-Mod tokamak plasma in a short pulse. (c) Temporal behavior of N atoms injected in a longer pulse. The numbers of impurity atoms are compared with (a) the integrated number of Ar atoms injected, and (b) and (c) the normalized injection rate of N. Reprinted from [13], Copyright (1997), with permission from Elsevier.

increase in the number of N atoms, here estimated by measuring the N VII emission line with a vacuum ultraviolet (VUV) spectrometer, in the plasma stops almost simultaneously with the end of the N gas puff. Thus, an evident exponential time decay of line emission from the highly ionized N is observed. In this case, the number of N atoms in the plasma is proportional to the injection rate.

Figure 4.10 shows an example discharge for an impurity transport study using an Ar gas puff in the KSTAR tokamak [14]. Here, to reduce the rise time of the increment of Ar atoms in the plasma, a short and intense Ar gas puff, corresponding to the injection of about 10^{18} Ar atoms, was performed in the stationary phase of a KSTAR H-mode discharge. As can be seen, even after the Ar gas puff, the plasma current, I_p, the line-integrated electron density, n_e, and the core electron temperature were not changed. However, the temporal evolution of the Ar^{15+} line emission, measured with a VUV spectrometer, increased rapidly and decreased gradually. Figures 4.10(b) and (c) show the plausible diffusivity and convection velocity profiles obtained by the SANCO impurity transport code. As can be easily recognized, the neoclassical diffusivity is much smaller than the diffusivity estimated by the SANCO code, and the neoclassical convection velocity is also much smaller than the

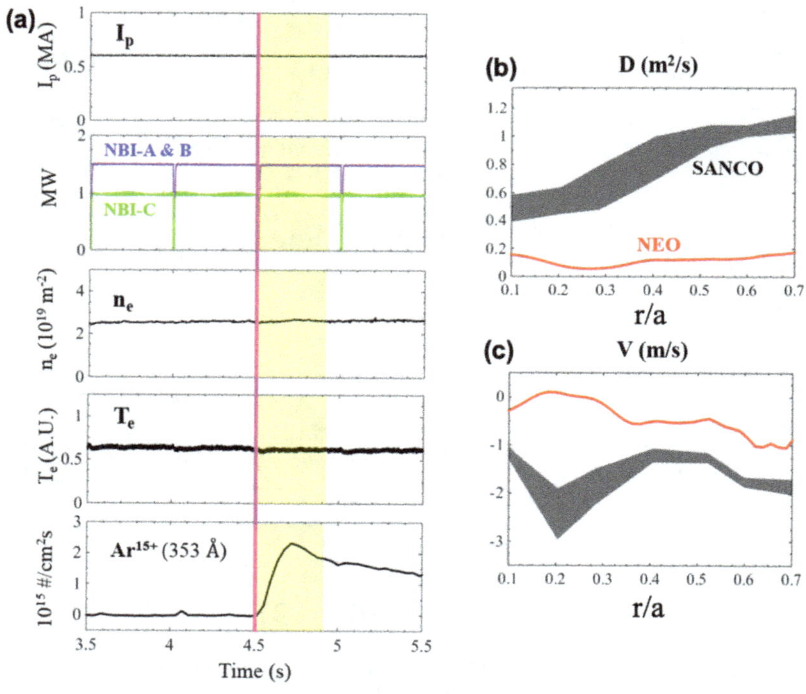

Figure 4.10. Example of experimental results of a gas-puff experiment for impurity transport study in the KSTAR tokamak. (a) Time traces of plasma current I_p, NBI heating power, line-integrated electron density n_e, core electron temperature T_e, and Ar^{15+} emission in KSTAR H-mode plasma. (b) and (c) Comparison of neoclassical D and V profiles (red solid lines) obtained by the NEO code calculation with those (gray areas) evaluated by the SANCO calculation. Reproduced from [14].

convection velocity estimated by the code. However, the sign of the convection velocity is almost the same (negative). Therefore, the experimental results cannot be explained only by neoclassical transport; therefore, contributions from turbulent transport should also be considered.

To obtain impurity transport coefficients such as diffusivity and convection velocity using a gas puff, a perturbation technique featuring a modulated gas puff has also been utilized, as in the case with a fuel gas such as a hydrogen isotope. Figure 4.11 shows experimental results for an impurity transport study using a modulated He gas puff in a reversed-shear plasma in the JT-60U tokamak [15]. In this discharge, to prolong the steady-state phase of the discharge, a feedback control for the neutral beam injection (NBI) heating power (P_{NBI}) using a neutron yield rate (S_n) signal was applied from $t = 4.6$ s. Following this, the He gas puff was

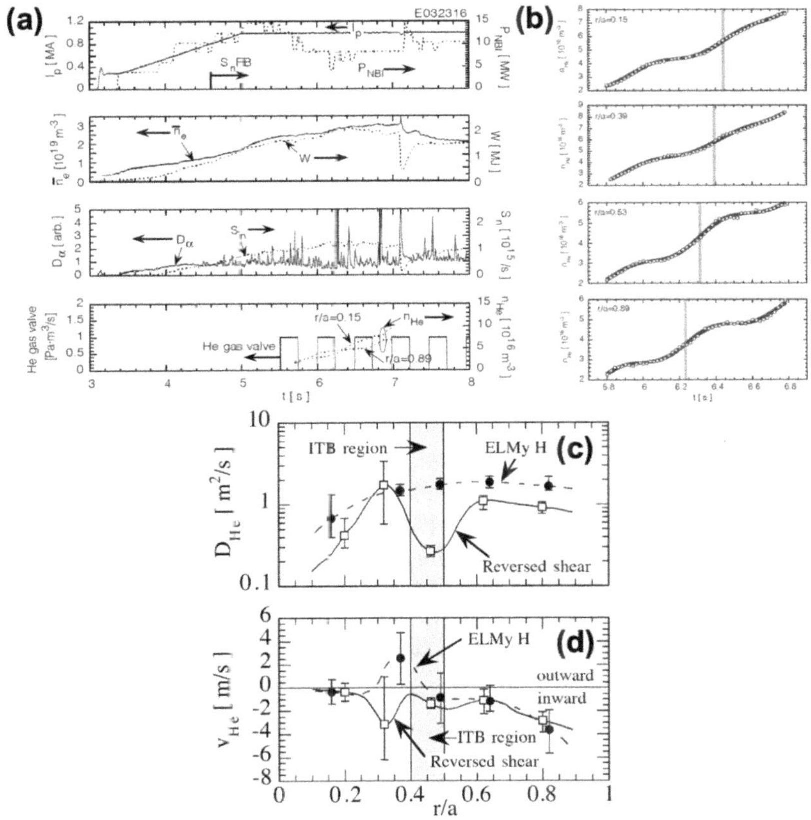

Figure 4.11. Example of experimental results of a gas-puff modulation experiment for impurity transport study in the JT-60U tokamak. (a) Time traces of typical plasma parameters, such as plasma current and line-averaged electron density and others, in a reversed-shear plasma of the JT-60U tokamak. (b) Temporal evolutions for helium (He) ion density measured with charge-exchange spectroscopy at different normalized minor radii. (c) Experimentally estimated profiles of diffusivity and convection velocity profiles for He ions. The shaded areas represent the ITB region. Reproduced from [15].

performed periodically at 2 Hz. Although the line-averaged electron density was slightly increased until the MHD collapse event occurred at $t = 7.1$ s, the line-averaged electron density was not modulated due to the He gas-puff modulation. However, as shown in figure 4.11(b), a modulation in the local He ion density, which can be measured using charge-exchange spectroscopy, was obtained due to the He gas-puff modulation. This is a crucial point that requires emphasis when dealing with impurity gas-puff modulation schemes in the study of impurity transport. The vertical line in figure 4.11(b) represents when the modulated component reaches zero. A phase delay toward the plasma center can be seen clearly in figure 4.11(b). When almost no delay in the phase toward the plasma center is observed, the He ions are considered to have penetrated quickly into the plasma core. This then indicates that the diffusivity for the He ions is also high. The impurity transport coefficients can be obtained using this perturbation analysis technique with a modulated gas puff even in the ITB region. Figures 4.11(c) and (d) show profiles of He transport coefficients (diffusivity coefficient and convection velocity) obtained using perturbation transport analysis. In reversed-shear JT-60U plasma, ITBs for electron density, electron temperature, and ion temperature are observed. An ITB for the He ion is also observed, and the ITB region for the He is indicated in gray in figures 4.11(c) and (d). It is observed that the He diffusivity in the ITB region is much lower than that outside the ITB region.

4.2.1.1 Impurity transport analysis with supersonic molecular beam injection

A supersonic molecular beam injection (SMBI) method has been developed to improve the location of the fuel delivery in the plasma and its efficiency as compared to the conventional gas-puff method, and is now widely used in many MCF devices. A supersonic molecular beam (SMB) is produced by adiabatic expansion of pressurized gas. The pressure around the beam ejection port should be as low as possible so as to eliminate collision effects with the surrounding gas. This is not a problem since the pressure is low enough in MCF devices. Molecules in the SMB travel directionally without colliding with each other. For the basic theory of a SMB and its experimental proof, see the studies of Kantrowitz & Grey and Kistiakowsky & Slichter [16, 17]. The beam's high directivity can create a localized particle source, an advantage in impurity transport analysis. Furthermore, its fast time response also can improve the abovementioned problem in gas-puffing methods; in other words, it can reduce recycling. The study of impurity transport using an SMBI has been performed in the HL-2A [18] and WEST (formerly Tore Supra) tokamaks [19]. Figure 4.12 shows experimental results of an impurity transport study using a nitrogen SMBI in the Tore Supra. For comparison, the N gas puff was also performed in a similar discharge. After the N gas puff, the UV line emission of N gradually increased, followed by a gradual increase in SXR intensity from the central chord. In the case of the N SMBI, on the other hand, the UV line emission of N increased and decreased rapidly almost immediately thereafter. An increase in SXR intensity from the central chord was also observed much earlier than in the N gas puff. These behaviors are quite similar to those obtained by another impurity injection method, a LBO (described in a later subsection). As shown in

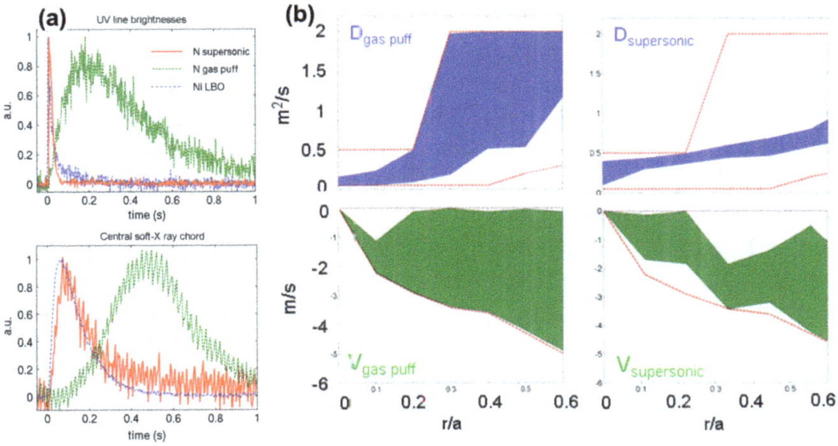

Figure 4.12. Example of experimental results of a supersonic molecular beam experiment for impurity transport study in the Tore Supra tokamak. (a) Comparison of temporal evolution between gas puff (green), supersonic molecular beam injection (SMBI, red), and laser blow-off (LBO, blue) in edge UV line emission (top) and core soft X-ray emission (bottom). (b) Radial profiles of diffusivity (top) and convection velocity (bottom) for nitrogen (N) gas puff (left) and N SMBI (right). The red dashed lines define the preset domain for the variation, and the color-hatched areas represent the estimated values including the uncertainties. Reproduced from [19].

figure 4.12(d), the N SMBI results provided a narrower range of impurity transport coefficients (diffusivity coefficient and convection velocity) than the N gas-puff results. It should be noted, however, that there are areas where the ranges do not overlap. This may be affected by ambiguity in the data analysis and uncertainty in the data obtained with measurements. In any case, the experimental results show that, as with the gas puff, the SMBI can be used to determine impurity transport coefficients with similar or better accuracy.

4.2.2 Impurity transport analysis with a laser blow-off

A charge and mass dependence of the impurity transport is also essential to fully understand the characteristics of impurity transport in magnetically confined plasmas. Highly charged (i.e., high-mass) impurity ions might be distributed inhomogeneously in the poloidal direction on a magnetic flux surface. To this end, to investigate the charge and mass dependence of impurity transport in magnetically confined plasmas, methods have been developed to intentionally inject elements other than those derived from the PFCs or those that can be introduced externally via the gas-puff method into the magnetically confined plasmas.

The LBO technique was developed as one such method to inject non-intrinsic impurities into magnetically confined plasmas. The LBO method is a rather old impurity injection method, developed in the mid-1970s, to overcome the problems inherent to using gas puffs for impurity transport studies in magnetically confined plasmas. The working principle of a LBO is as follows. A transparent glass plate is coated with the impurities to be injected into the plasma. Then, the coated side of the transparent glass is placed as the plasma-facing side and irradiated with a high-power laser from its back. The interaction with the high-power laser vaporizes the impurity film, and atoms, molecules, clusters, ions, electrons, etc., are explosively ejected from its surface. However, because the magnetic field confines the plasma exits, ions and electrons do not reach the plasma, and atoms, molecules, and clusters of impurities are injected into the plasma. The amount of impurities ejected can be varied according to the energy density of the laser.

The main advantage of the LBO method is the ability to inject non-gaseous, non-intrinsic impurities into the plasma at a relatively high speed. Compared to gaseous impurities at room temperature, impurities injected by the LBO method can be injected at relatively high speed, i.e., with higher energy, and thus can be injected into the plasma beyond the scrape-off layer region at the periphery. In addition, because the impurities injected by the LBO method are non-gaseous, recycling can be reliably suppressed. Therefore, the entire process, from the point when the impurities are injected into the plasma until the impurities leave the plasma, can be accurately measured, i.e., the impurity confinement time can be accurately estimated. The high directivity and fast time response of the LBO method, as in the case of SMBI shown earlier, also contribute significantly to improving the problems of impurity injection via the gas-puffing method. The glass plates can be coated with various impurities, such as lithium fluoride (LiF), aluminum (Al), calcium oxide (CaO), scandium (Sc), Ti, chromium (Cr), Fe, Ni, molybdenum (Mo), tungsten (W), etc., as single elements or as compounds using multi-layer films or other techniques. This makes it possible to study the transport of various impurities in magnetically confined plasmas.

Figure 4.13 shows an example discharge for an impurity transport study using a Mo LBO in the Frascati Tokamak Upgrade (FTU) tokamak [20]. As can be seen in figure 4.13(a), when the Mo impurity was injected into the plasma via the LBO, the SXR intensities measured at different radii immediately increased. At the same time, the central electron density and temperature were not varied. This is an essential feature of the LBO method. The LBO method can inject impurities without causing perturbations to the electron density and temperature in the core plasma. Therefore, the LBO method allows investigation of the transport of impurities in a trace limit ($\alpha = n_Z Z^2 / n_i \ll 1$) in non-perturbed plasmas. After the Mo injection by LBO, the SXR intensity at $r = 5\,\text{cm}$ is higher than that at $r = 0\,\text{cm}$. As can be seen in figures 4.13(b) and (c), reconstruction of the radial profile of the experimentally observed SXR intensity at the maximum SXR intensity yields a hollow shape with a peak at $r = 7\,\text{cm}$. After that, the SXR intensity at the plasma center increased while the peak position of the SXR intensity slightly moved toward the plasma center. Figures 4.13(c) and (d) show the Mo transport coefficients (diffusion coefficient and convection velocity) calculated to reproduce the experimental result. It can be seen that outward convection velocity inside $r = 10\,\text{cm}$ and strong inward convection velocity outside $r = 10\,\text{cm}$ must be considered to maintain the shape of the experimental SXR intensity profile.

One of the advantages of the LBO method is the ability to inject a variety of non-intrinsic impurities into the magnetically confined plasma. This advantage allows for

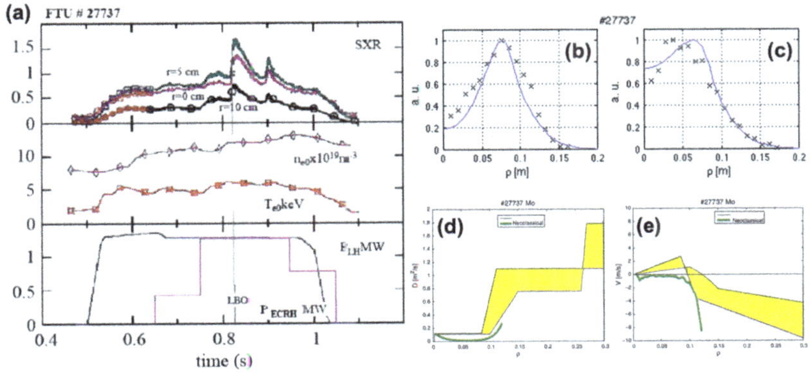

Figure 4.13. Example of experimental results of a laser blow-off (LBO) experiment for impurity transport study in the FTU tokamak. (a) Time traces of soft X-ray (SXR) intensities measured at different radii, core electron density and temperature, and input heating power with LBO injection. Also shown are the radial profiles of experimentally obtained and calculated SXR intensities (b) at the time of the maximum SXR intensity, $t = 0.83$ s, and (c) at the decay phase, $t = 0.86$ s, as well as profiles of the (d) diffusivity and (e) convection velocity obtained experimentally (the uncertainty of the estimation is indicated by the yellow colored zone) and those obtained with the calculation based on neoclassical theory. Reproduced from [20].

a more systematic study of the Z dependence of impurity transport in magnetically confined plasmas. Figure 4.14 shows the results of one such more systematic study of the Z dependence of impurity transport in plasmas of the W7-X stellarator, taking advantage of this LBO method [21]. As can be seen in figure 4.14, various impurities, from the low-Z silicon (Si) to the high-Z W, were injected into W7-X plasma by the LBO method, and subsequently the time evolution of the line-emission intensity from each impurity ion in the X-ray region were measured by a high-resolution crystal imaging spectrometer (HR-XIS). As figure 4.14(b) shows, the decay time of the line-emission intensity from each impurity ion, i.e., the impurity confinement time, increases very gradually with increasing impurity atomic number. This indicates that turbulent transport dominates the impurity transport in W7-X plasmas heated at high electron cyclotron heating power.

4.2.3 Impurity transport analysis with an impurity pellet

The impurity pellet injection method was originally developed for another plasma diagnostic purpose rather than for the study of impurity transport. The injection of low-Z materials such as Li, B, and C into plasma was first considered for utilization in the study of alpha-particle physics [22] and the internal magnetic field structure [23], for the control of wall conditions [24] and disruption [25]. Naturally, the impurities injected were also utilized to study the impurity transport in magnetically confined plasmas. In fusion research up until around 1990, the PFCs in MCF devices were mainly made of metals such as Cr, Fe, and Ni. Therefore, at that time, the injection of low-Z impurities was also used for the study of non-intrinsic impurity transport. It was then found that metallic impurities mixed into the plasma

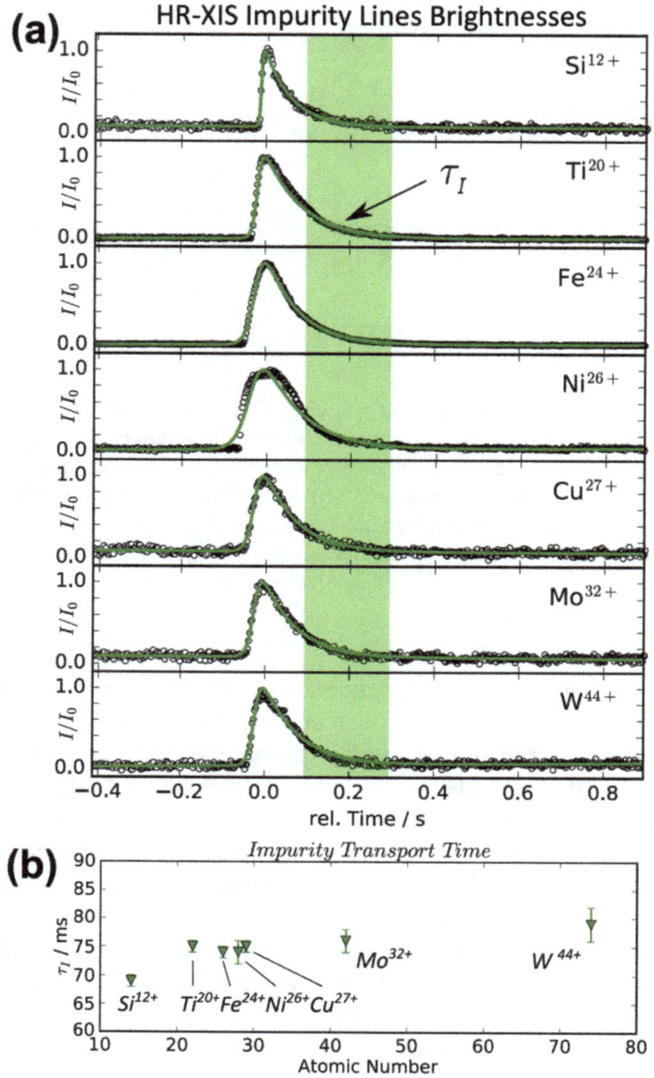

Figure 4.14. Examples of various kinds of impurity injection with LBO scheme in a MCF device, the W7-X stellarator. (a) Time traces for the normalized brightness from seven different kinds of impurity ion after LBO. The decay time (corresponding to the impurity transport time, T_I) of the normalized brightness was estimated in the shaded area. (b) Experimentally obtained impurity transport time as a function of the atomic number of each impurity ion. Reproduced from [21] CC BY 4.0.

became a barrier to improved plasma performance due to, for example, impurity accumulation in the plasma center, and their presence thus posed a problem. Therefore, beginning around 1990, in fusion research most plasma-facing materials in MCF devices were changed from metals to graphite (C). In the current era of research, metal impurities such as Al, Fe, and Ni have also been injected via the impurity pellet method to study non-intrinsic impurity transport.

The main technique for injecting an impurity pellet into a plasma is usually the pipe-gun method. A pipe gun is a gun that ejects projectiles using the energy of compressed air or other pressurized gases. In magnetically confined plasmas, He, the lightest and safest gas possible, is commonly used as the propellant gas to maximize pellet velocity. In addition, spherical or cylindrical pellets made entirely of a specific element or compound are used as projectiles. Simply put, the velocity of the pellet depends on the inner diameter and length of the barrel in the impurity pellet injector when the diameter of the cross-sectional area of the pellet and the inner diameter of the barrel are approximately the same. Thus, if the cross-sectional diameter of the pellet is relatively small compared to the inner diameter of the barrel, the pellet will not be sufficiently accelerated by the gas. Even in such a case, the problem can be solved using a container called a sabot. By using the pellet container, the pellet can be sufficiently accelerated by the propellant gas. However, in this case it is necessary to add a mechanism such as a sabot stopper to the impurity pellet injector so that only the pellet in the sabot is injected into the plasma.

The impurity pellet injection method also has all the advantages of non-intrinsic impurity injection. It is possible to inject the impurity pellet almost precisely at a scheduled time while the magnetically confined plasma is sustained. Usually, the timing of the propellant gas ejection is set in experiments to accelerate the impurity pellet. The time it takes for the impurity pellet to travel from the impurity pellet standby position to the plasma edge depends on the distance from the impurity pellet standby position to the plasma edge and the impurity pellet velocity. The velocity of the impurity pellet will almost be the same if the cross-sectional diameter of the impurity pellet and the pressure of the propellant gas are the same, although there is some variation. Therefore, if preliminary experiments are conducted and a database of impurity pellet velocities is compiled, the impurity pellet can be injected into the magnetically confined plasma at an approximately scheduled time. The impurity to be injected can be selected from various options, but some limitations exist. The main limitations consist in that the impurity must be a stable material which can be safely handled at room temperature, and that it must be capable of being shaped into an appropriate size in a pellet form. However, if an aforementioned sabot is used, it may be possible that there are no limitations to this shaping.

The impurity pellets, after entering the plasma, interact with the surrounding plasma, ablate, become ionized, and assimilate with the surrounding plasma. This process continues until the impurity pellets are entirely ablated. Thus, the radial profile of the impurity deposited by the impurity pellet extends from where the impurity pellet begins to penetrate the plasma to where it finishes ablating completely. The broad radial profile of the impurity deposition is often problematic in impurity transport studies. For example, when the spatial structure of the impurity transport varies within the impurity deposition profile, the information regarding the impurity transport that can be estimated might be then averaged over the width of the deposition profile. Figure 4.15 shows an example discharge for an impurity transport study using a pellet injection of Li in an RFX-mod reversed-field pinch device [26]. After the Li pellet was injected into the plasma, at 185 ms, as can be seen in the third frame from the top of figure 4.15(a), the line-averaged electron

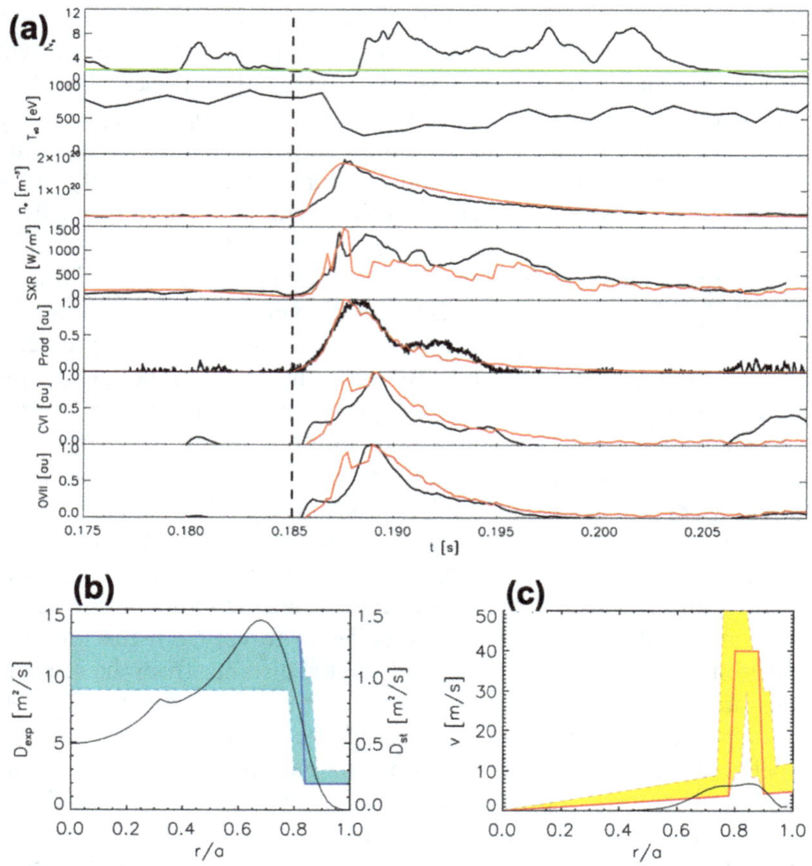

Figure 4.15. Example of experimental results of an impurity pellet injection experiment for impurity transport study in a MCF device, an RFX-mod reversed-field pinch device (a) Time traces of typical plasma parameters, such as line-averaged electron density and core electron temperature and others around the lithium (Li) pellet injection, the time of which is indicated with the vertical dashed line. (b) Transport coefficient profiles for Li ion determined by the experiment. the uncertainties of the estimation are indicated by the colored areas. The black lines are obtained by the calculation. Reproduced from [26].

density drastically increased in the order of 10^{20} m^{-3}. This is due to the relatively large size of the Li pellets (cylinders of $l \sim 5$ mm, $r = 0.75$ mm), as they are routinely utilized for wall conditioning. This large density perturbation provoked an increment of the radiation from the plasma and a decrease in the central electron temperature. The experimentally measured plasma parameters were compared with those calculated with simulation codes used to estimate the impurity transport coefficients. The radial profiles of the experimentally estimated Li transport coefficients with uncertainty ranges are shown in figures 4.15(b) and (c), with respective colored areas. The calculated Li transport coefficients are also shown in figures 4.15(b) and (c), with black lines. From these figures, the diffusion coefficient over the is found to be much higher than that calculated, and a large convection velocity region, which is well above that calculated, is also found to exist. It should be noted here that the perturbation (relatively significant in this case) due to the impurity pellet injection may be a disadvantage in the study of impurity transport in magnetically confined plasmas. This is because, especially in this case, the impurities will likely be above the trace limit ($\alpha \ll 1$), and the impurities themselves may affect impurity transport.

There is a new-type impurity pellet that overcomes the disadvantages of the conventional impurity pellet. A tracer-encapsulated solid pellet (TESPEL), which has been developed at National Institute for Fusion Science (NIFS) in Japan, is, simply stated, a double-layered impurity pellet. Usually, the outer layer, the so-called 'shell', of the TESPEL is made of a polystyrene $(C_8H_8)_n$ polymer, and the tracer impurity is embedded in the core. Because of this structure, when a TESPEL is injected into plasma, the polystyrene shell first ablates, ionizes, and then assimilates into the surrounding plasma. After the shell of the TESPEL finishes ablating, the inner tracer impurities ablate, ionize, and are assimilated into the surrounding plasma. A TESPEL has four essential characteristics: (i) direct local deposition of tracer impurities in plasma is possible, (ii) the amount of tracer impurity deposited in the plasma can be accurately determined in advance, (iii) a relatively wide range of tracer materials can be selected, (iv) the flexible size of the TESPEL allows for variable tracer deposition locations. The first feature is unique, even considering the other impurity injection methods discussed above. This feature allows a TESPEL to inject tracer impurities into three-dimensional isolated areas with pinpoint accuracy. The second feature is also unique to TESPEL. TESPELs are fabricated by hand. When fabricating a TESPEL, the size of each impurity micro-particle to be encapsulated is measured. In this way, when using a TESPEL, the impurities injected into the plasma can be determined in advance.

Figure 4.16 shows an example discharge for an impurity transport study using a TESPEL in the LHD heliotron [27]. As shown in figure 4.16(a), for electron density, the TESPEL was injected into plasma with an IDB and, thus, a highly peaked electron density profile was formed. In this experiment, Ti micro-particles were embedded in the TESPEL. To experimentally estimate the deposition location of the tracer impurity, the ablation emissions from the shell and inner core of the TESPEL were measured, respectively. Figure 4.16(b) shows radial profiles of the Hα emission (blue solid line) from the shell and the Ti I emission (green solid line) from the inner core measured through the corresponding interference filter. The Ti I emission seems to appear even from the edge ($\rho \sim 1$) of the plasma. Of course, this is not correct. That emission is continuum emission from a dense cold plasma surrounding the ablating TESPEL, which can also be measured through the Ti I filter. Around $\rho = 0.45$, the Ti I emission intensity is still high, but the Hα emission is seen to decrease sharply. The sharp decrease in Hα emission indicates that the ablation of the outer layer of the TESPEL is ending. At the same time, the Ti I emission intensity remains high, suggesting that the tracer impurity from the inner core has been exposed to the surrounding plasma and has begun to ablate. This suggests the Ti tracer impurity is locally deposited between $\rho = 0.4$ and $\rho = 0.5$, well inside the

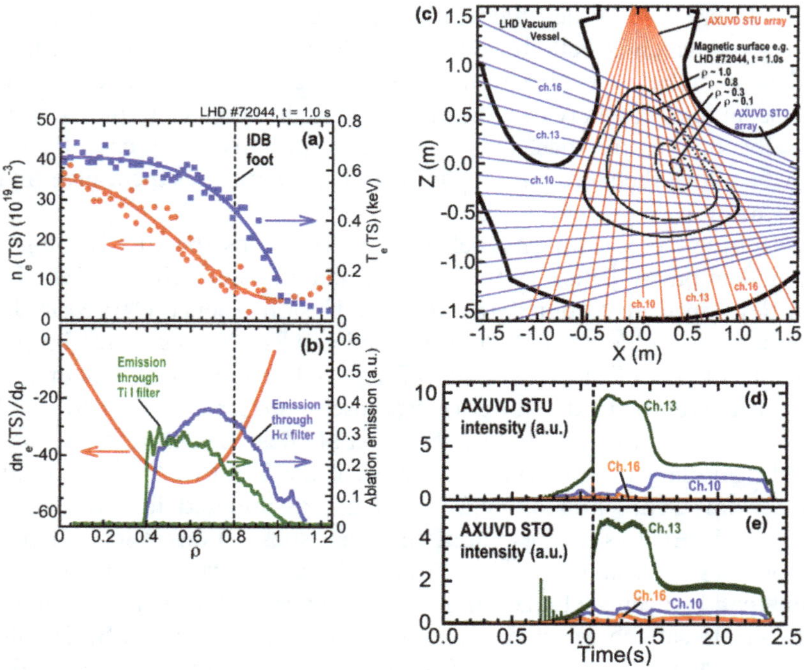

Figure 4.16. Example of unique experimental results of an injection experiment utilizing an advanced impurity pellet, tracer-encapsulated solid pellet (TESPEL) for impurity transport study in a MCF device, the LHD heliotron. The TESPEL was injected into a very-high-density LHD plasma with an internal diffusion barrier (IDB). (a) Electron density (red) and temperature (blue) profiles just before the TESPEL injection, (b) TESPEL ablation emissions measured through Hα (blue) and Ti I (green) interference filters. The vertical dashed line indicates the location of the IDB. (c) LHD magnetic flux surfaces corresponding to the time just before the TESPEL injection, including the lines of sight of the diode array (AXUVD) measurements from the upper port and from the horizontal port. Temporal evolutions of the AXUVD signal intensities from the upper port (d) and from the horizontal port (e). Reprinted from [27], with permission from The Japan Society of Plasma Science and Nuclear Fusion Research.

IDB region. As can be seen in figures 4.16(d) and (e), the signal intensities of channel (ch.) 13 of the absolute extreme ultraviolet photodiode (AXUVD) array at the upper port (AXUVD STU) and that at the horizontal port (AXUVD STO) after the TESPEL injection are highly increased, and are sustained until around t = 1.4 s. This behavior suggests that the Ti tracer impurity is deposited locally in the core region of the LHD plasma and is kept there. The Ti tracer impurity deposited in the LHD core plasma is immediately affected by the transport characteristics there, which verifies the unique advantage of the TESPEL technique.

References

[1] Edwards A W, Campbell D J, Engelhardt W W, Fahrbach H U, Gill R D and Granetz R S *et al* 1986 Rapid collapse of a plasma sawtooth oscillation in the JET tokamak *Phys. Rev. Lett.* **57** 210–3

[2] Ida K, Fonck R J, Sesnic S, Hulse R A and LeBlanc B 1987 Observation of z-dependent impurity accumulation in the PBX tokamak *Phys. Rev. Lett.* **58** 116–9

[3] Petrasso R, Seguin F H, Loter N G, Marmar E and Rice J 1982 Fully ionized and total silicon abundances in the Alcator-C tokamak *Phys. Rev. Lett.* **49** 1826–9

[4] Seguin F H, Petrasso R and Marmar E S 1983 Effects of internal disruptions on impurity transport in tokamaks *Phys. Rev. Lett.* **51** 455–8

[5] Nave M F F, Rapp J, Bolzonella T, Dux R, Mantsinen M J and Budny R *et al* 2003 Role of sawtooth in avoiding impurity accumulation and maintaining good confinement in JET radiative mantle discharges *Nucl. Fusion* **43** 1204–13

[6] Dux R, Peeters A G, Gude A, Kallenbach A and Neu RASDEX Upgrade Team 1999 Z-dependence of the core impurity transport in ASDEX Upgrade H mode discharges *Nucl. Fusion* **39** 1509–22

[7] Nunes I, Conway G D, Loarte A, Manso M, Serra F and Suttrop W *et al* 2004 Characterization of the density profile collapse of type I ELMs in ASDEX Upgrade with high temporal and spatial resolution reflectometry *Nucl. Fusion* **44** 883–91

[8] Putterich T, Dux R, Janzer M A and McDermott R M 2011 ELM flushing and impurity transport in the H-mode edge barrier in ASDEX Upgrade *J. Nucl. Mater.* **415** S334–9

[9] Perry M E, Brooks N H, Content D A, Hulse R A, Mahdavi M A and Moos H W 1991 Impurity transport during the H-mode in DIII-D *Nucl. Fusion* **31** 1859–75

[10] Ida K, Sakamoto Y, Yoshinuma M, Takenaga H, Nagaoka K and Hayashi N *et al* 2009 Dynamics of ion internal transport barrier in LHD heliotron and JT-60U tokamak plasmas *Nucl. Fusion* **49** 095024

[11] Rowan W L, Bespamyatnov I O and Fiore C L 2008 Light impurity transport at an internal transport barrier in Alcator C-Mod *Nucl. Fusion* **48** 105005

[12] Dux R, Giroud C and Zastrow K DJET EFDA Contributors 2004 Impurity transport in internal transport barrier discharges on JET *Nucl. Fusion* **44** 260

[13] McCracken G M, Granetz R S, Lipschultz B, Labombard B, Bombarda F and Goetz J A *et al* 1997 Screening of recycling and non-recycling impurities in the Alcator C-Mod tokamak *J. Nucl. Mater.* **241** 777–81

[14] Hong J, Henderson S S, Kim K, Seon C R, Song I and Lee H Y *et al* 2017 Modification of argon impurity transport by electron cyclotron heating in KSTAR H-mode plasmas *Nucl. Fusion* **57** 036028

[15] Takenaga H, Nagashima K, Sakasai A, Asakura N, Shimizu K and Kubo H *et al* 1999 Particle confinement and transport in JT-60U *Nucl. Fusion* **39** 1917–28

[16] Kantrowitz A and Grey J 1951 A high intensity source for the molecular beam. Part I. Theoretical *Rev. Sci. Instrum.* **22** 328–32

[17] Kistiakowsky G B and Slichter W P 1951 A high intensity source for the molecular beam. Part II. Experimental *Rev. Sci. Instrum.* **22** 333–7

[18] Cui X W, Cui Z Y and Feng B B 2013 Investigation of impurity transport using supersonic molecular beam injected neon in HL-2A ECRH plasma *Chinese Phys.* B **22** 125201

[19] Guirlet R, Villegas D, Parisot T, Bourdelle C, Garbet X and Imbeaux F *et al* 2009 Anomalous transport of light and heavy impurities in Tore Supra ohmic, weakly sawtoothing plasmas *Nucl. Fusion* **49** 055007

[20] Leigheb M, Romanelli M, Gabellieri L, Carraro L, Mattioli M and Mazzotta C *et al* 2007 Molybdenum transport in high density FTU plasmas with strong radio frequency electron heating *Plasma Phys. Control. Fusion* **49** 1897–912

[21] Langenberg A, Wegner T, Pablant N A, Marchuk O, Geiger B and Tamura N *et al* 2020 Charge-state independent anomalous transport for a wide range of different impurity species observed at Wendelstein 7-X *Phys. Plasmas* **27** 052510

[22] Fisher R K, Leffler J S, Howald A M and Parks P B 1988 Fast alpha diagnostics using pellet injection *Fus. Technol.* **13** 536–42

[23] Marmar E S, Terry J L, Lipschultz B and Rice J E 1989 Measurement of the current density profile in the Alcator C tokamak using lithium pellets *Rev. Sci. Instrum.* **12;60** 3739–43

[24] Strachan J D, Mansfield D K, Bell M G, Collins J, Ernst D and Hill K *et al* 1994 Wall conditioning experiments on TFTR using impurity pellet injection *J. Nucl. Mater.* **217** 145–53

[25] Pautasso G, Buchl K, Fuchs J C, Gruber O, Herrmann A and Lackner K *et al* 1996 Use of impurity pellets to control energy dissipation during disruption *Nucl. Fusion* **36** 1291

[26] Barbui T, Carraro L, Franz P, Innocente P, Munaretto S and Spizzo G 2014 Light impurity transport studies with solid pellet injections in the RFX-mod reversed-field pinch *Plasma Phys. Control. Fusion* **57** 025006

[27] Tamura N, Sudo S, Suzuki C and Funaba H 2015 Creation of impurity source inside plasmas with various types of tracer-encapsulated solid pellet *Plasma Fusion Res.* **10** 1402056

Chapter 5

Impurity transport across magnetic flux surfaces

The differences between impurity transport across the magnetic flux surfaces in tokamak and helical plasmas is described in this chapter. In helical plasma, the radial electric field has a significant impact on impurity transport, while the radial electric field effect is relatively weak in tokamak. In general, impurities tend to accumulate in the plasma center due to the inward convection in tokamak devices, especially in a good confinement regime. In contrast, in helical systems, the outward convection results in a hollow impurity profile, which can become extremely hollow (known as an impurity hole) in plasmas with an internal transport barrier.

5.1 Effect of radial electric field on impurity transport

Outward convection, due to positive radial electric field inside the last closed-flux surface (LCFS) and friction force in stochastic regions with open magnetic field, plays an important role in avoiding the accumulation of impurities in helical plasmas [2]. Inward convention in the plasma results in impurity accumulation, where the impurity concentration and radiation power in the core region increases in time during the steady-state phase. In contrast, the outward convection prevents impurity accumulation, and the impurity concentration and radiation power in the core region do not increase and are kept constant in time. In the LHD, it was found that the behavior of impurities (accumulation or shielding) is sensitive to the collisionality of the plasma and the radial electric field near the plasma edge. As can be seen in figure 5.1, impurity accumulation is observed in the plasma with a higher density of ~4×10^{19} m^{-3}, while there is no impurity accumulation observed in the plasma with a lower density of ~1×10^{19} m^{-3}. At higher density, the edge radial electric field at $\rho = 0.95$ is negative (ion root) and there is a significant increase of radiation power in time, which is clear evidence of impurity accumulation. In contrast, at lower density, the edge radial electric field is slightly positive (electron root) and there is no increase of radiation power in time, which is clear evidence of impurity shielding. In the electron-root regime, the positive radial

doi:10.1088/978-0-7503-1451-0ch5

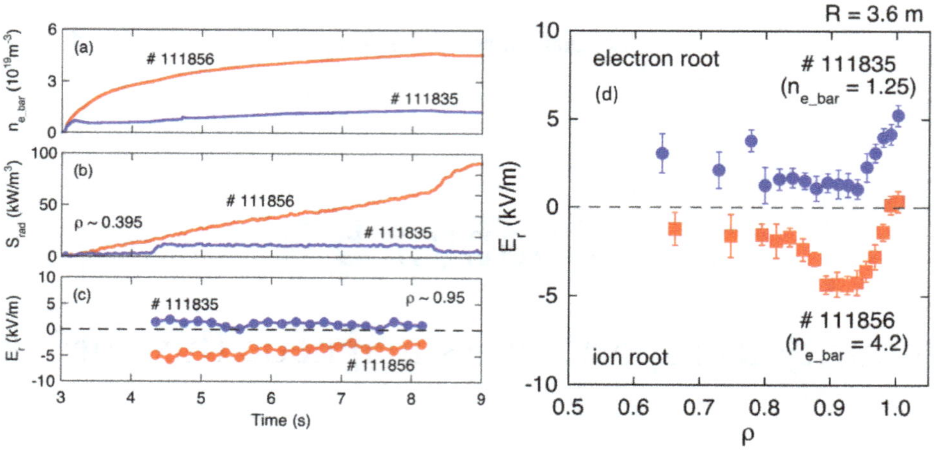

Figure 5.1. Time evolution of (a) line-averaged electron density, (b) radiation power density in the central region, and (c) radial electric field near the plasma edge, as well as (d) radial profile of radial electric field for low-density (electron root) and high-density (ion root) plasma in the LHD. Reproduced from [1].

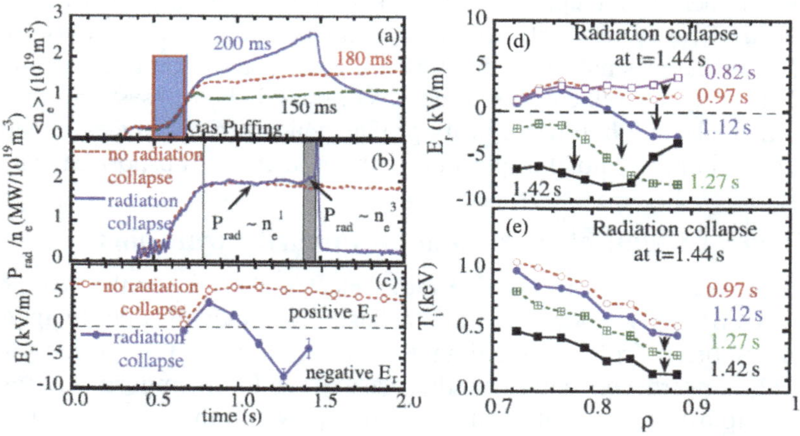

Figure 5.2. Time evolution of (a) electron density in a discharge with a 150 ms, 180 ms, and 200 ms neon (Ne) gas puff, (b) radiation power normalized by density, and (c) radial electric field at $\rho = 0.89$ in a discharge with a 180 ms and 200 ms Ne gas puff, as well as radial profiles of (d) radial electric field and (e) ion temperature. The discharge is terminated by radiation collapse for the 200 ms Ne gas puff, while there is no radiation collapse for the 150 ms and 180 ms Ne puffs. Reproduced from [3].

electric field increases toward the plasma edge ($\rho = 1$), which has a significant role in shielding of the impurity influx, which is referred to as E_r shielding.

The important role of the edge radial electric field in radiation collapse was studied by applying a neon (Ne) gas puff with different amounts, which is proportional to Ne puff pulse width, near the threshold for radiation collapse. Figure 5.2 shows the time evolution of electron density and radiation power normalized by density and the radial electric field near the plasma edge at $\rho = 0.89$. When the Ne puff pulse width is below the threshold (150 ms and 180 ms), there is no radiation collapse and the edge radial electric field remains positive. However, when the Ne puff pulse width is above the threshold (200 ms), the edge radial electric field changes its sign from positive to negative and radiation collapse starts to occur. As seen in the figures 5.2(d) and (e), the change in the radial electric field starts from the plasma edge, and the negative electric field extends to the plasma core. This change in the radial electric field is due to the decrease of temperature (as seen in the radial profile of the ion temperature), which starts from the plasma edge and propagates inside in a timescale of 0.3 s. This experiment demonstrates the important feedback process in radiation collapse. In this feedback process, an increase of radiation power causes a decrease of temperature and increase of collisionality. As the collisionality increases, the edge radial electric field becomes more negative, and this negative radial electric field enhances the impurity influx and radiation power. Due to this feedback process, radial collapse takes place when the amount of neon and radiation power at the plasma edge exceeds the threshold values.

Figure 5.3 shows the plasma parameter space of electron density, n_e, and electron temperature, T_e, at the plasma edge in LHD helical plasma for discharges with impurity accumulation and impurity shielding [4]. The dashed and solid lines indicate the boundaries of the parameter space for impurity accumulation. In the parameter space with a higher temperature than the dashed boundary, the positive radial electric field enhances the outward convention of the impurity transport, which is the aforementioned E_r shielding. This E_r shielding in the core of electron-root plasma is due to the outward convection predicted by the neoclassical radial flux of impurities. In contrast, in the parameter space with a lower temperature than the solid boundary, the electrostatic force and the friction force toward the divertor plate overcome the thermal force toward the core plasma, which is known as impurity shielding in the ergodic layer. This impurity shielding effect in the ergodic layer due to the electrostatic and friction forces becomes stronger as the edge density is increased. The open red points represent discharges with a strong suppression of impurity accumulation.

Similar experimental results were observed in the Wendelstein 7-AS stellarator [5]. As the electron density is increased, the radial electric field becomes more negative, as predicted by neoclassical theory. As can be seen in figure 5.4, a significant increase of impurity confinement time is observed as the electron density is increased from 5×10^{18} m^{-3} to 1×10^{20} m^{-3}. The increase of impurity confinement time is more than two orders of magnitude, while the increase of energy confinement time is only one order of magnitude (roughly proportional density). This is due to a change of sign of the convection of the impurity near the magnetic axis. As seen in figure 5.4(b), the diffusion coefficient decreases as the electron density is increased from a medium level to a high level. Figure 5.4(c) shows that the outward convection velocity (positive values) near the plasma center disappears, while the inward convection velocity (negative values) near the plasma edge is unchanged. The dark shaded region in figure 5.4(a) is the high-density H-mode (HDH) regime [6], where the energy confinement time is doubled by the transition

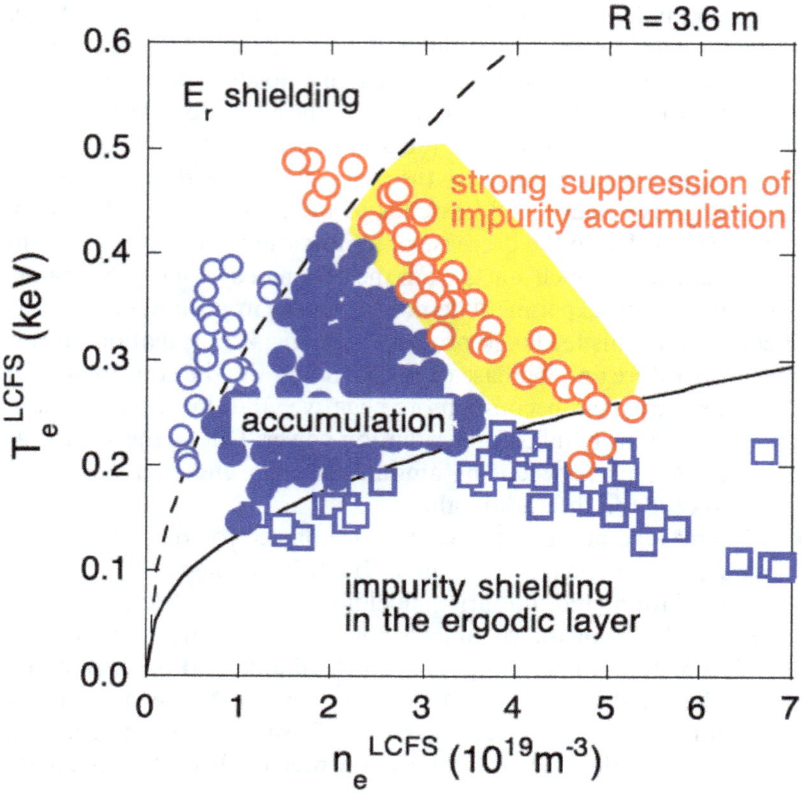

Figure 5.3. Impurity behavior (accumulation or shielding) in the plasma parameter space of electron density n_e and electron temperature T_e at the plasma edge in LHD helical plasma. The blue closed and open symbols indicate discharges with and without impurity accumulation, respectively. Reproduced from [4].

from a low-confinement mode (L-mode) to a high-confinement mode (H-mode). In the HDH discharge, quasi-stationary discharges with line-averaged densities of up to 4×10^{20} m^{-3}, radiation levels up to 90%, and partial plasma detachment at the divertor target plates can be simultaneously realized. It should be noted that the impurity confinement time decreases one order of magnitude, while the energy confinement increases at the transition. This is in contrast to the increase in both energy confinement time and impurity confinement time at the transition from L-mode to H-mode in tokamak devices. Therefore, the HDH regime is considered to be a more preferable condition for fusion reactors.

The experiments both in the LHD and Wendelstein 7-AS devices demonstrate the importance of the radial electric field in impurity transport. It is also interesting to note that there is an impurity accumulation window in the density regimes of helical plasmas. At lower density, the positive radial electric field inside the LCFS plays an important role in reducing the impurity influx (namely, E_r shielding). As the density increases, this E_r shielding disappears and impurity accumulation begins. However,

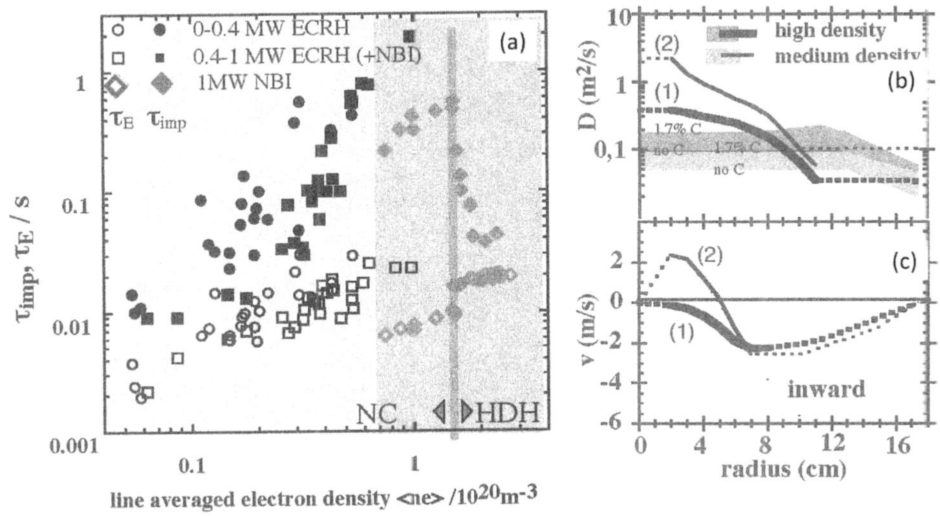

Figure 5.4. (a) Energy and impurity confinement times vs electron density in normal confinement (NC) electron cyclotron resonance heating (ECRH) and neutral beam injection (NBI) plasma, and during the high-density H-mode (HDH) transition (dark shaded region), as well as radial profiles of (b) diffusion coefficient and (d) convection velocity in medium-density and high-density plasma. [5] Taylor & Francis Ltd. http://tandfonline.com.

when the density is increased further, impurity shielding outside the LCFS (in the ergodic layer) begins to play a role in preventing the influx of impurities into the plasma. Since the physical mechanisms to determine impurity transport are quite different inside and outside the LCFS, impurity transport inside and outside the LCFS has been studied separately with different transport simulation codes. Although determining the impurity transport across the LCFS is important to predict the total fraction of impurity in the plasma, the physics for impurity transport across the LCFS has not been sufficiently studied. Impurity transport across the LCFS is discussed in section 7.3.

5.2 Impurity accumulation

As already discussed in previous chapters, the accumulation of impurities toward the plasma center in magnetically confined plasmas is one of the long-standing, serious problems regarding impurities in fusion research. As already mentioned in chapter 1.1, if impurity accumulation occurs in the plasma center, low-Z impurities can cause fuel (bulk hydrogen isotopes) dilution and high-Z impurities can cause significant radiation losses due to line emissions from partially ionized high-Z impurities.

Ignition conditions for deuterium–tritium (or DT) fusion reactors are obtained from plasma heating by α (high-energy helium) particles, which must adequately compensate for plasma energy loss due to transport and radiation. Figure 5.5 shows ignition condition curves calculated when considering contamination from helium (He) and tungsten (W). From this figure, the minimum $nT\tau_E$ required at a given

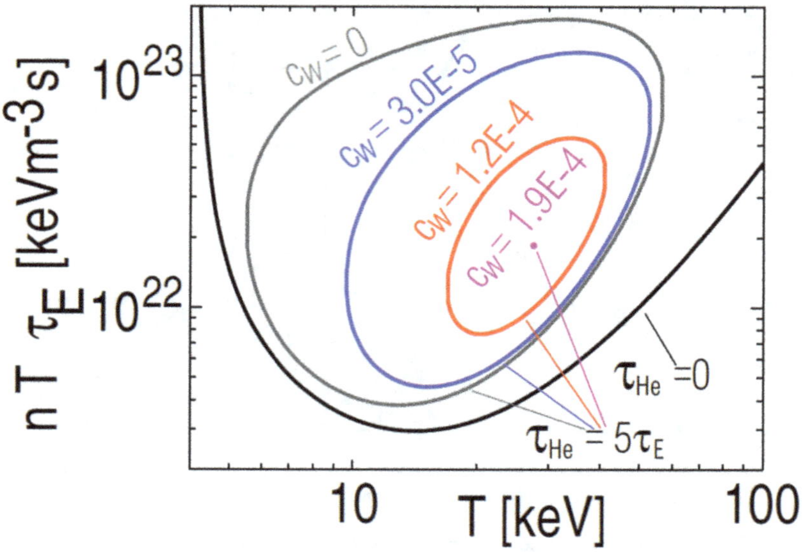

Figure 5.5. Ignition curves in the presence of different tungsten W concentrations for $\tau_{He} = 5\tau_E$, including curves without He and without W for reference. Reproduced from [7]. © International Atomic Energy Agency. Published by IOP Publishing. All rights reserved.

temperature can be assessed. Here, the plasma energy loss due to transport is considered by the global energy confinement time τ_E. Since the He, a byproduct of the fusion reaction, can be exhausted by a pumping system via a divertor, here He is considered in terms of confinement time. For a detailed scheme with which to evaluate ignition curves, readers are referred to the study of Reiter *et al* [8]. A He confinement time of $\tau_{He} = 0$ is an ideal case, but it should be emphasized that there is still an explicit boundary on the ignition condition. The He confinement time $\tau_{He} = 5 \times \tau_E$ is predicted based on the ITER reference scenario [9], corresponding to a core He concentration below 6%. As shown in figure 5.5, the ignition condition shifts to the higher-temperature side as the W concentration increases. When the W concentration exceeds 1.9×10^{-4}, there is no longer any condition that allows ignition. In a real fusion reactor, impurities other than He and W are known to exist in the plasma, so a lower W concentration (possibly around several 10^{-5}) must be maintained to sustain nuclear burning. This implies that W (i.e., high-Z impurities) accumulation toward the plasma center must be avoided as much as possible.

5.2.1 Impurity accumulation in tokamak plasmas

The characteristics of impurity accumulation differ slightly between tokamak and helical plasmas. Figure 5.6 shows examples of various profiles of JT-60U tokamak plasmas with internal transport barriers (ITBs) [10]. Figure 5.6(a) is for a reversed-shear configuration, and figure 5.6(b) is for a high-β_p mode. In the reversed-shear case, the ITB causes steep gradients in the electron and ion temperature and electron density profiles. The temperature and density profiles are flattened inside the

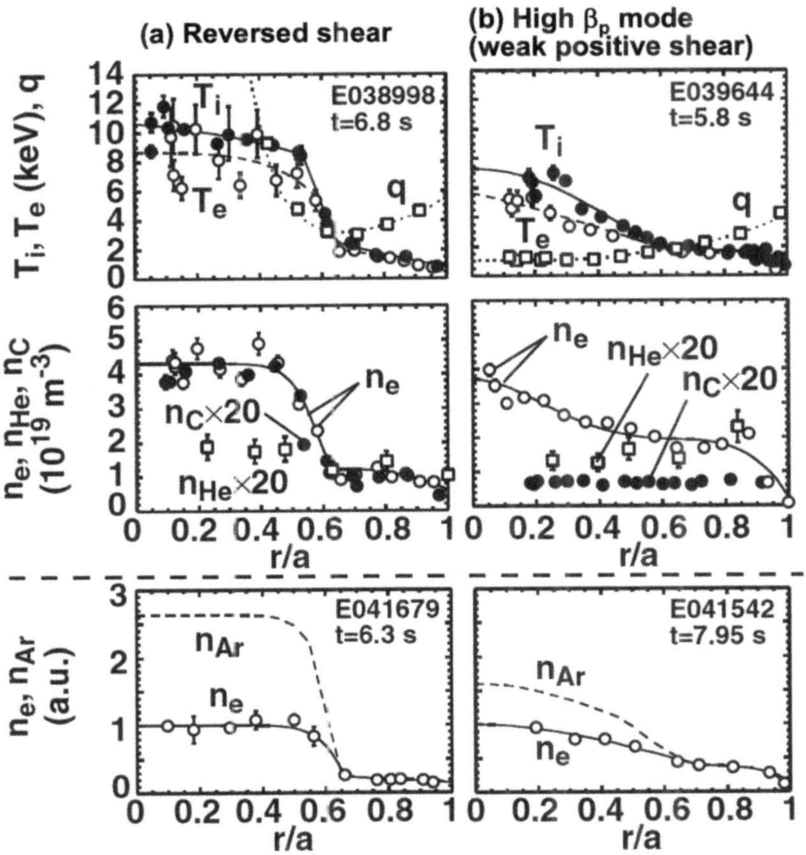

Figure 5.6. Top panels: Profiles of ion temperature T_i (solid circle), electron temperature T_e (open circle), and safety factor q (open square). Middle panels: Density profiles of electron n_e (open circle), helium n_{He} (open square), carbon n_C (solid circle). Bottom: Profiles of n_e (open circle) and argon density n_{Ar} (dashed line). Left figures were obtained in reversed-shear JT-60U plasma, and right figures were obtained in high-β_p mode (weak positive shear) JT-60U plasma. Reproduced from [10]. © International Atomic Energy Agency. Published by IOP Publishing. All rights reserved.

location where the steep gradient is generated. The He density profile is generally flat, while the carbon (C) density profile has almost the same shape as the electron density profile. The argon (Ar) density profile is more strongly peaked than the electron density profile. On the other hand, in the high-β_p mode, the electron and ion temperature profiles and the electron density profile inside the foot point of the ITB increase gently. In this case, the He and C density profiles have a flat overall shape. In contrast, the Ar density profile has a slightly peaked shape compared to the electron density profile. In tokamak plasmas, the transport of low-Z impurities in the plasma core is dominated by turbulent transport, with neoclassical transport having little impact. Since the turbulent transport has no component that drives the inward impurity flow, the accumulation of low-Z impurities does not occur. On the other

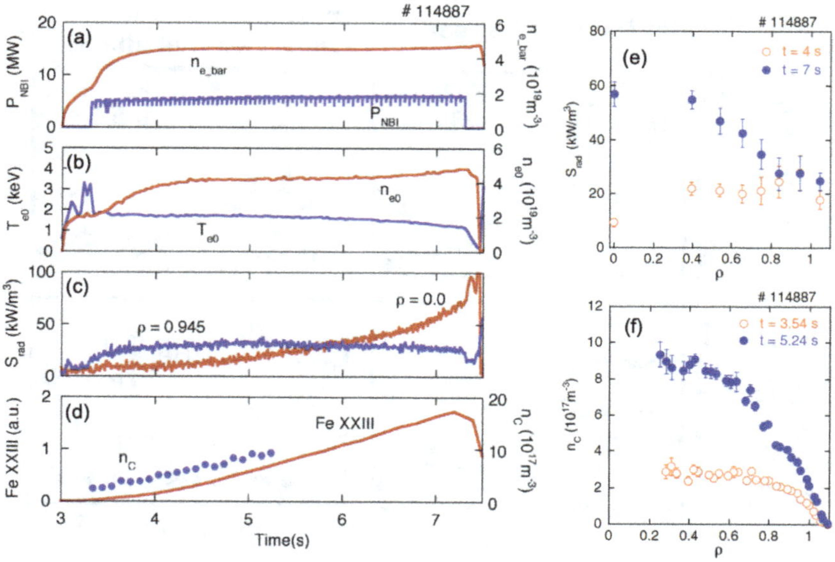

Figure 5.7. Time evolution of (a) the line-averaged electron density n_{e_bar} (red) and injected neutral beam power P_{NBI} (blue), (b) the central electron density n_{e0} (red) and electron temperature T_{e0} (blue), (c) the radiated power density S_{rad} at the plasma center ($\rho = 0.0$) and at the plasma edge ($\rho = 0.945$), and (d) the Fe XXIII line emission intensity and central carbon (C) density. Also shown are radial profiles of (e) radiated power density and (f) C density n_{C} at different times. Reprinted from [11], Copyright (2017), with permission from Elsevier.

hand, in the transport of high-Z impurities in the plasma core, neoclassical transport is dominant because the turbulent transport condition is relatively minor and the contribution of neoclassical transport is more significant. Therefore, as figure 5.6 shows, in the core region of tokamak plasmas, the higher Z the impurities, the stronger likelihood that accumulation will occur.

5.2.2 Impurity accumulation in helical plasmas

On the other hand, in helical plasmas, impurity accumulation can occur both for low-Z and high-Z impurities. Figure 5.7 shows an example discharge exhibiting impurity accumulation in the LHD heliotron [11]. As figure 5.7(c) shows, the plasma radiation from the plasma core continues to increase until the end of the discharge, which suggests that an accumulation of impurities took place in this discharge. Figure 5.7(d) shows that the C density in the plasma core (although data is available only until the middle of the discharge) and the line emission intensity of Fe XXIII also continue to increase with time. Figures 5.7(e) and (f) also show that the radiated intensity and the radial profile of the C density become more peaked with time compared to those at the initial phase of the discharge. However, the electron density profile remained flattened when the peaked C density distribution was obtained. These experimental results in the LHD show that when high-Z impurities accumulate, low-Z impurities also accumulate. This is because the convection terms of neoclassical transport, especially the radial electric field, play a significant role in impurity transport in helical plasmas. Although the inward convection velocity due to the negative radial electric field has a linear dependence on the Z number, the inward convection also affects the low-Z impurities.

The world's largest stellarator, the Wendelstein 7-X (W7-X) in Germany, which has recently commenced experiments, is designed to reduce (in other words, optimize) the level of neoclassical transport by tailoring the magnetic configuration, which has been confirmed experimentally [12]. It should be noted here that the W7-X does not have a magnetic configuration design that includes a reduction of turbulent transport. Therefore, in the W7-X, no accumulation of high-Z impurities occurs even in plasmas with high electron density (high collisionality), where impurity accumulation is expected, i.e., where the radial electric field is negative [13]. On the other hand, in W7-X plasmas where the turbulent transport is small or suppressed, the contribution of neoclassical transport is relatively significant, and impurity accumulation is observed. In the LHD, as shown in figure 5.3, impurity accumulation may not occur even in an operation regime where impurity accumulation is expected from the point of view of neoclassical transport, i.e., the sign of the radial electric field. This is considered to be partly due to increased turbulent transport. Therefore, in helical plasmas, the competition between neoclassical and turbulent transport is also an essential determinant of impurity transport.

5.3 Poloidal asymmetry of impurity density

In the framework of standard neoclassical transport theory for magnetically confined toroidal plasmas, it is generally accepted that the impurity densities, along with other bulk ion and electron densities, are uniformly distributed on the magnetic flux surface. This is because of the rapid parallel transport that occurs along the magnetic field lines. Nonetheless, from the earliest stages of experimentation, it has been noted that the impurity densities display a variety of asymmetries on the magnetic flux surface, specifically on the poloidal surface. These asymmetries take the form of up and down (up–down) and in and out (in–out). While this result could be easily anticipated if the impurity particle source had a highly localized and strongly poloidal asymmetrical impurity particle flux, most observations are not contingent on the source's localization. In other words, the mechanism that generates the poloidal asymmetries exists as an impurity transport parallel to the magnetic field lines. It should be noted that almost no bulk ion or electron density inhomogeneity on the magnetic flux surface has been observed. The following subsections detail the various poloidal asymmetries of impurity densities observed in magnetically confined toroidal plasmas.

5.3.1 In–Out/Out–In asymmetry of impurity density

5.3.1.1 Inertia (centrifugal) force

In toroidally rotating tokamak plasmas, a concentration of heavy impurity ions at the low-field side (LFS), in other words, an out–in asymmetry of impurity ions, has been observed. The first observation of such a concentration of impurity ions at the LFS was made using an ultra-soft X-ray tomography technique in the neutral beam injection (NBI)-heated H-mode plasma of the ASDEX tokamak [14]. Figure 5.8 shows example discharges exhibiting the impurity concentration at the LFS of the magnetic flux surfaces in the ASDEX Upgrade [15] and JET [16] tokamaks. As shown in figure 5.8, a region of high soft X-ray (SXR) emissivity with a banana-like shape is clearly formed on the LFS of the tokamak plasmas. Such out–in asymmetry of the impurity ions is more pronounced for impurities with higher Z. In the ASDEX Upgrade tokamak, there is an example where such out–in asymmetry of impurity ions was not observed in argon or neon, but rather in krypton [17]. A centrifugal force due to toroidal rotation is the mechanism that produces such an effect. Due to centrifugal force, the density asymmetry of impurity species z becomes of order ϵM_z^2, where $\epsilon = r/R$ represents the inverse aspect ratio and M_z represents the Mach number of the impurity species of interest. M_z^2 can be expressed with

$$M_z^2 = \frac{m_z \omega_z^2 R^2}{2T_z}, \tag{5.1}$$

where the impurity mass is m_z, ω is the toroidal angular rotation frequency of the impurities, R is the major radius, and the impurity temperature is denoted by T_z. Since the impurity temperature is usually assumed to be the same as the bulk ion temperature, T_i (i.e., $T_z \sim T_i$), the asymmetry of the impurity density becomes of order $M_i^2(m_z/m_i)$. Therefore, high-Z impurities can become concentrated at the LFS of tokamaks even with moderate plasma rotation velocities, and low-Z impurities may be concentrated at the LFS of strongly rotating tokamak plasmas.

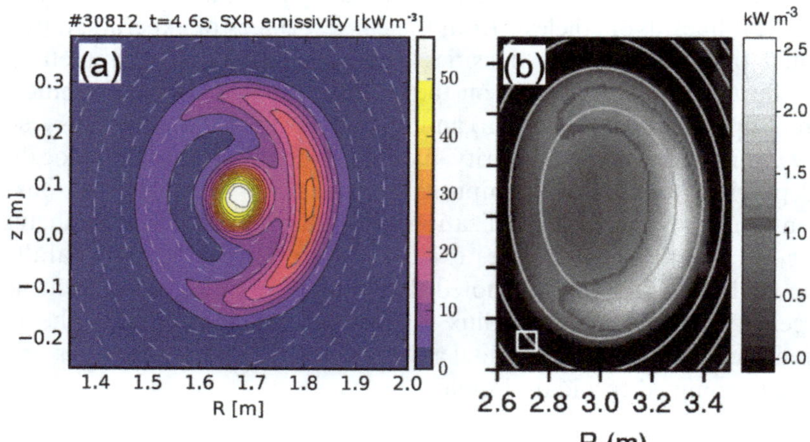

Figure 5.8. Tomographic reconstructed images of the soft X-ray (SXR) emissivity obtained in (a) ASDEX Upgrade tokamak discharge (#30812) at a time of 4.4 s with 2.5 MW NBI heating only, and (b) JET tokamak discharge (#40551) at a time of 6.72 s (514 ms after a nickel laser blow-off) with 17 MW NBI and 6 MW radio-frequency heating powers. Panel (a) reproduced from [15]. © 2017 Max-Planck-Institut fur Plasmaphysik. Published by IOP Publishing. All rights reserved. Panel (b) reprinted from [16], with the permission of AIP Publishing.

Neoclassical transport of heavy impurity ions can be significantly modified when a poloidal asymmetry of the heavy impurity ions occurs. Recent theoretical studies have extended and modified the neoclassical impurity transport model for when a poloidal asymmetry of the heavy impurity ions takes place in tokamak plasmas. When trace impurities are deeply in the Pfirsch–Schlüter (PS) regime and bulk ions are in the banana regime, the neoclassical impurity particle flux obtained by this series of studies can be summarized as [18]

$$\frac{R\langle\Gamma_Z^{\rm neo}\rangle}{\langle n_z\rangle} = q^2 D_c Z\left[\left(\frac{1}{Z}\frac{R}{L_{nZ}} - \frac{R}{L_{ni}} + \frac{1}{2}\frac{R}{L_{Ti}}\right)P_A - 0.33 P_B f_c \frac{R}{L_{Ti}}\right], \tag{5.2}$$

where two geometrical factors, P_A and P_B, are formulated with the following equations:

$$P_A = \frac{1}{2\epsilon^2}\frac{\langle B^2\rangle}{\langle n_Z\rangle}\left[\left\langle\frac{n_Z}{B^2}\right\rangle - \left\langle\frac{B^2}{n_Z}\right\rangle^{-1}\right]^2, \quad P_B = \frac{1}{2\epsilon^2}\frac{\langle B^2\rangle}{\langle n_Z\rangle}\left[\frac{\langle n_Z\rangle}{\langle B^2\rangle} - \left\langle\frac{B^2}{n_Z}\right\rangle^{-1}\right]^2. \tag{5.3}$$

In equation (5.2), q is the charge, $D_c = \rho_i^2/\tau_{ii}$ is the classical diffusion coefficient with the magnetic-flux-surface-averaged ion Larmor radius squared, $\rho_i^2 = v_{thi}^2/\langle\Omega_i^2\rangle$, and the ion–ion collision time, $\tau_{ii} = 3(2\pi)^{3/2}\epsilon_0^2 m_i^{1/2} T_i^{3/2}/(n_i e^4 ln\Lambda)$; R/L_{nZ}, R/L_{ni}, and R/L_{Ti} denote the normalized gradient of impurity ions, bulk ion, and ion temperature, respectively; f_c is the fraction of circulating (non-trapped) particles; and $\langle\rangle$ denotes the flux surface average. In equation (5.2), the conventional neoclassical diffusion coefficient, pinch, and temperature screening are multiplied by the factor of P_A. Standard neoclassical impurity transport is recovered with factors of $P_A = 1$ and $P_B = 0$ (for the case of impurity ions with a poloidal symmetry and in a large aspect ratio limit). On the other hand, the case of an ultimately strong poloidal asymmetry of impurity ions can be expressed with $P_A = P_B = 1/2\epsilon^2$. Here, as can be easily seen, the term multiplied by the factor of P_B in equation (5.2) can relatively mitigate the temperature screening effect. Taking into account the contribution of the factor of P_A, a strong poloidal asymmetry of impurities can cause an enhancement of neoclassical transport as well as a reduction of the temperature screening effect.

Figure 5.9 shows an excellent example of how neoclassical impurity transport, taking into account a strong poloidal asymmetry in the impurities, reproduces the experimental results observed in the JET tokamak [18]. The experimental data shown in figure 5.9 (red solid lines) were obtained by a SXR camera diagnostic, which was converted to tungsten density as a function of the major radius along with the horizontal axis at the level of the magnetic axis for JET pulse #83351. At the time of 9.34 s (left frame of figure 5.9), a hollow electron density profile and peaked electron temperature profile were observed. Then, at 10.42 s (middle frame of figure 5.9), the electron density profile changed from hollow to flat, and the central electron temperature decreased as the tungsten ions increased at the plasma center, as shown in the figure. At 12.37 s (right frame of figure 5.9), after more time had elapsed, the accumulation of tungsten ions in the plasma center progressed, and the electron density profile became more peaked. Here, to reconstruct the estimated

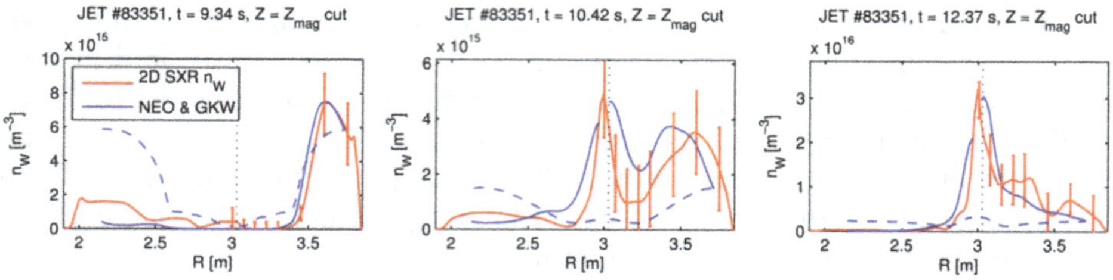

Figure 5.9. Three time slices, (a) 9.32 s, (b) 10.42 s, and (c) 12.37 s, of the estimated tungsten density (red solid lines) as a function of the major radius along the horizontal axis at the level of the magnetic axis for JET pulse #83351 from a soft X-ray (SXR) camera diagnostic. Each frame includes the calculated tungsten density by the NEO and GKW codes with (blue solid lines) and without (blue dashed lines) the inclusion of the centrifugal effect. Reprinted from [18], with the permission of AIP Publishing.

tungsten density, numerical analysis was performed with the drift kinetic neoclassical code NEO, including the contribution of turbulent tungsten transport, which was computed with the gyrokinetic code GKW. Calculations were performed in two cases, one including the effect of centrifugal force and the other not, to judge the impact of centrifugal force. Details of the calculations are given in Angioni *et al* [18]. At 9.34 s, both calculations reproduced the highly hollow tungsten density profile well, except for the calculation result on the high-field side (HFS), which excluded the centrifugal force effect. On the other hand, at 10.42 s and 12.37 s, the calculations with the centrifugal force effect quantitatively reproduced the concentration of tungsten ions toward the plasma center. In contrast, the calculations without the centrifugal force effect did not reproduce the same phenomenon. These results indicate that a poloidal asymmetry of the impurity ions in tokamak plasmas significantly impacts the impurity transport itself (e.g., impurity accumulation) and must be considered and predicted appropriately.

In helical plasmas, the viscosity in the toroidal direction is higher than in tokamak plasmas [19]. Therefore, the toroidal rotation velocity of the plasma is lower comparatively than in that of tokamak plasmas, and in–out asymmetry of impurity ions derived from the centrifugal force due to toroidal rotation has never been observed before.

5.3.1.2 Electrostatic force

In tokamak plasmas, it has been noted that a particular phenomenon occurs whereby heavy impurity ions become concentrated on the HFS. This asymmetry between inside and outside impurity ions has been consistently observed, particularly when ion cyclotron resonance heating (ICRH) dominates. Such a HFS concentration of heavy impurity ions was first observed in nickel laser blow-off experiments in the JET tokamak with ICRH only [16]. Figure 5.10 shows example discharges exhibiting the impurity concentration at the HFS of the magnetic flux surfaces in the ASDEX Upgrade [15] and JET [20] tokamaks. As can be seen in figure 5.10 (and more clearly in figure 5.10(a), showing the results from the ASDEX Upgrade tokamak), there is a distinct formation of a region that exhibits high SXR emissivity on the HFS of the tokamak plasmas. What is noteworthy in this case is

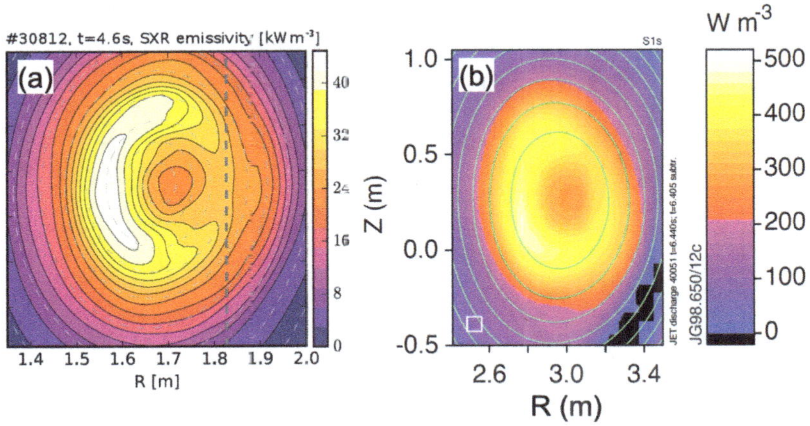

Figure 5.10. Tomographic reconstructed images of the soft X-ray (SXR) emissivity obtained in (a) tASDEX Upgrade tokamak discharge (#30812) at a time of 4.6 s with 2.5 MW NBI and 4.3 MW ICRH powers, and (b) JET tokamak pulse (#40551) at a time of 6.44 s (35 ms after a nickel laser blow-off) with 3 MW radio-frequency heating power. In (a), the vertical dashed line indicates the ion cyclotron resonance. Panel (a) reproduced from [15]. © 2017 Max-Planck-Institut fur Plasmaphysik Published by IOP Publishing. All rights reserved. Panel (b) reproduced from [20]. © IOP Publishing Ltd. All rights reserved.

that figure 5.10(a) corresponds to the same discharge as figure 5.8(a); the only difference is that of the addition of ICRH after the time shown in figure 5.8(a). This implies that the impurities, concentrated on the LFS, moved to the HFS due to a driving force created by the ICRH. The origin of this driving force is believed to be the poloidal asymmetry of the electrostatic potential. ICRH can sustain poloidal variation of the electrostatic potential on the flux surface by magnetically trapped minority ions being selectively heated by ICRH. The magnetically trapped minority ions concentrated on the LFS can push the heavy (and highly charged) impurity ions toward the HFS. These highly charged impurity ions are sensitive to even small potential fluctuations, and they respond in a Boltzmann-like manner with $n_z(\theta)/\langle n_z\rangle \sim \exp(-Ze\Phi(\theta)/T_z)$, where Φ is the electrostatic potential and θ is the poloidal angle. If solely the inertial (centrifugal) force and electrostatic potential hold significance in parallel force balance, then the form of the poloidal variation of any thermalized ion species, be it the bulk ion or impurity, shall follow

$$n_a(\theta) = n_{aR0} \exp\left(-\frac{Ze\Phi(\theta)}{T_a} + \frac{m_a\omega_a^2(R(\theta)^2 - R_0^2)}{2T_a}\right), \tag{5.4}$$

where R_0 is the major radius at the LFS, and n_{aR0} is the density of the ion species at R_0 [21]. Since the radio-frequency (RF)-heated ion species can be modeled with a bi-Maxwellian distribution characterized by a temperature anisotropy ($T_\perp$, $T_\parallel$), equation (5.4) can be deformed as [22]

$$n(\theta) = n_{R0}\frac{T_{\perp(\theta)}}{T_{\perp R0}} \exp\left(-\frac{eZ\Phi(\theta)}{T_\parallel} + \frac{m\omega^2(R(\theta)^2 - R_0^2)}{2T_\parallel}\right), \tag{5.5}$$

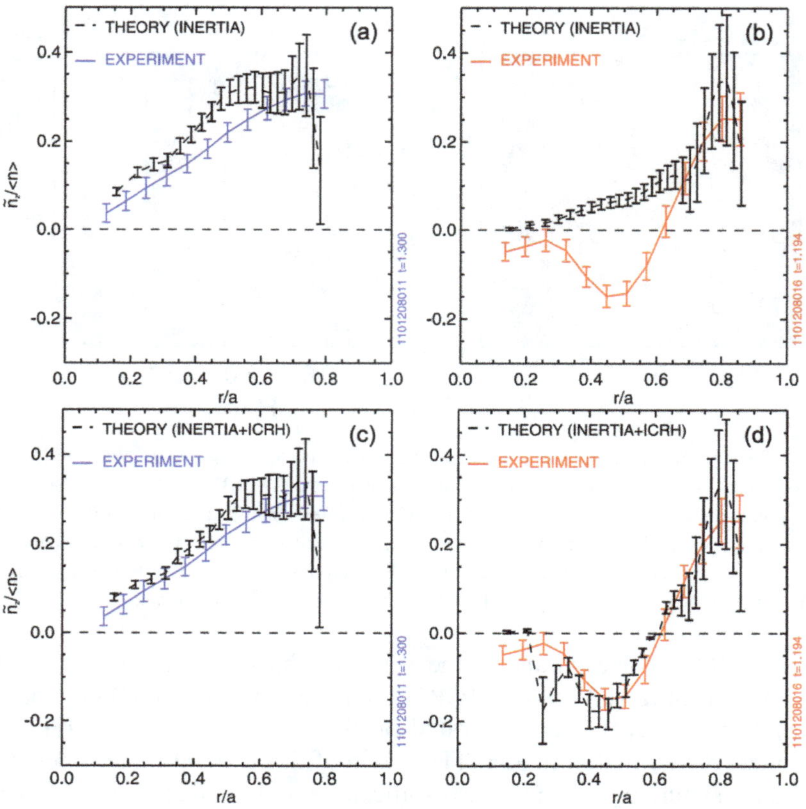

Figure 5.11. Comparison of the measured (solid) and calculated (dashed black) impurity density asymmetries for Alcator C-MOD tokamak experiments. (a) and (c) Radio-frequency (RF)-heated plasmas on the HFS with B_t = 4.4 T, and (b) and (d) RF-heated plasmas on the LFS with B_t = 6.0 T. In the calculation of (a) and (b), only an inertia (centrifugal force) is considered, and in (c) and (d) a combined effect of the centrifugal force and ICRH is considered. Reproduced from [21].

where

$$\frac{T_{\perp(\theta)}}{T_{\perp R0}} = \left[\frac{T_{\perp 0}}{T_{\parallel}} + \left(1 - \frac{T_{\perp R0}}{T_{\parallel}}\right)\frac{B_{R0}}{B(\theta)}\right]^{-1}. \tag{5.6}$$

Equation (5.5) is also valid for the isotropic case with $T_{\perp}/T_{\parallel} = T_{\perp(\theta)}/T_{\perp R0} = 1$. Since temperature anisotropy is the essence here, magnetically trapped electrons (due to ECRH) and ions (due to NBI) may also cause the poloidal potential variation and contribute to parallel impurity transport. In the Alcator C-Mod tokamak, a strong asymmetry of molybdenum density on the flux surface has been observed routinely in plasmas heated by a hydrogen (H) minority ICRH scheme. Figure 5.11 shows comparison results between the measured asymmetries of the impurtiy density and the calculated ones using the above-described analytical model [21]. When ICRH is applied to the HFS, there is no large discrepancy between the measured and calculated values in the region of $r/a < 0.7$. If anything, as shown in figure 5.11(a), the discrepancy between the measured and calculated values appears to be slightly larger overall when the temperature anisotropy due to ICRH is not considered. On the other hand, when ICRH is applied to the LFS, as shown in figure 5.11(b), the measured and calculated values differ significantly between $0.3 < r/a < 0.65$, even though the measured and calculated values agree well at $r/a \sim 0.8$ when temperature anisotropy due to ICRH is not considered. The region with $0.3 < r/a < 0.65$ coincides with the location where the heating power applied by the ICRH is expected to be absorbed. In contrast, as shown in figure 5.11(d), when the temperature anisotropy due to ICRH is considered, the measured and calculated values are in good quantitative agreement, indicating that the model presented above is able to explain this experimental result.

In non-axisymmetric and non-quasi-symmetric magnetic configurations, the neoclassical transport processes exclusively determine the radial electric field. The ion and electron particle fluxes should be equivalent across a magnetic flux surface to establish the radial electric field. This is quite different from tokamaks. However, toroidal plasmas can have a natural mechanism for electric fields parallel to the flux surface as well as the mechanism with the toroidal rotation and magnetically trapped energetic particles. In tokamaks, such an electric field is relatively small. But, in stellarators, such a poloidally (and also toroidally) varying component of the electrostatic potential can be significantly large enough so as to affect the cross-field impurity transport. Poloidal variation of the electrostatic potential is related to the radial width of the ion trajectories, as discussed in the studies of Mynick and Ho & Kulsrud [23, 24]. Furthermore, the impact of the poloidal variation of the electrostatic potential, denoted by Φ_1, on the radial impurity transport has been studied in a number of theoretical and numerical studies. However, no direct effect of Φ_1 on impurity transport in the radial direction has been experimentally observed, although there is an example experimental observation of magnetic flux surface inhomogeneity of the electrostatic potential at the plasma periphery in the TJ-II stellarator [25]. Recently, the impact of the poloidally varying electrostatic potential invoked by magnetically trapped ions due to ICRH was studied numerically. Still, its effect on the impurity transport was established to be small in the case considered [26].

5.3.2 Up–Down/Down–Up asymmetry of impurity density

The impurity parallel force balance can be formulated as

$$\frac{m_a n_a \omega^2}{2} \nabla_{\parallel} R^2 + eZ_a n_a \nabla_{\parallel} \Phi + T_a \nabla_{\parallel} n_a = R_{a,\parallel}. \tag{5.7}$$

The first term, $(m_a n_a \omega^2/2)\nabla_{\parallel} R^2$, is the inertia (cetrifugal force), and the second term, $eZ_a n_a \nabla_{\parallel} \Phi$, is the electrostatic force, which have already been discussed in section 5.3.1. The third term, $T_a \nabla_{\parallel} n_a$, is the pressure term, the poloidally varying components of which is to be all in n_a. This pressure term is dominant for light impurities.

The friction force, $R_{a,\parallel}$, is only relevant at the plasma edge, where large gradients in ion density and temperature are expected with high collisionality.

Up–down asymmetries of impurity line emission have been observed from the earliest stages of tokamak experiments, for instance with a vertically scanning spectrometer. In tokamak experiments, the up–down asymmetry was observed before the in–out asymmetry of the impurity line emission because it is easier to measure from the horizontal diagnostic port than from the vertical diagnostic port when considering the observation of the whole poloidal cross-section of the tokamak plasma. Figure 5.12 shows an example discharge of up–down asymmetry of impurity line emission observed with the vertically scanned UV spectrometer in the PDX tokamak [27]. As shown in figure 5.12, the up–down asymmetry of the C III and C V emission lines was reversed when the ∇B drift direction was reversed. To compare the experimental results and the neoclassical transport model, the sight-line-integrated line emission intensity, ϵ, is modeled with $\epsilon(r, \theta) = \epsilon_0(r)(1 + \eta \sin \theta)$, where η is the asymmetry parameter. As a result, the observed asymmetry of C V line emission can be reproduced with the neoclassical model when the main ions are assumed to be in the banana regime. As discussed in Brau *et al* [27], the impact of the neutral C and H and the anomalous radial diffusion on the vertical asymmetry of C V line emission was excluded. The up–down asymmetry of impurity radiation has also been observed in stellarator plasmas. Figure 5.13 shows an example discharge with up–down asymmetry of plasma radiation observed with the bolometer camera system in the W7-X stellarator [28]. As can be seen in figure 5.13, the radiation region from the impurities was concentrated in the lower-left side of the figure, obtained with the standard 5/5 magnetic island configuration in the normal toroidal field direction (SDC+). When the toroidal field direction was reversed (with SDC-), the intense radiation region was moved to the upper-left side of the figure. The edge-localized two-dimensional (2D) radiation pattern, which is inverted according to the toroidal field direction, was captured by an SXR camera as well as the bolometer

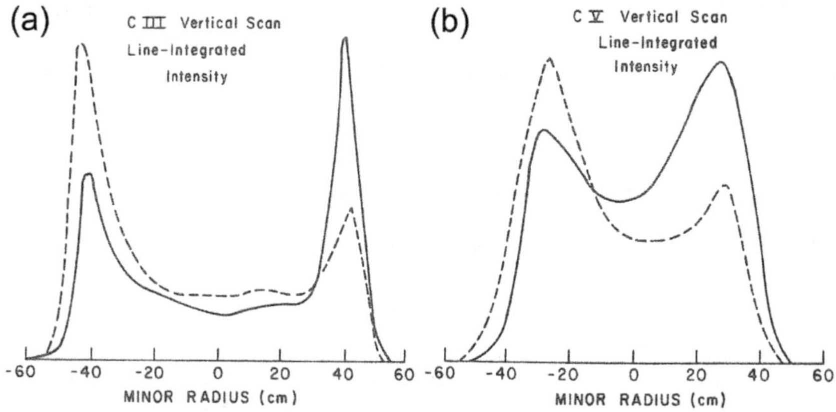

Figure 5.12. Vertical radial profiles of (a) C III and (b) C V emissions obtained by scanning the UV spectrometer vertically on a shot-to-shot basis. The solid (dashed) line corresponds to a vertical ion drift directed to downward (upward). Reproduced from [27].

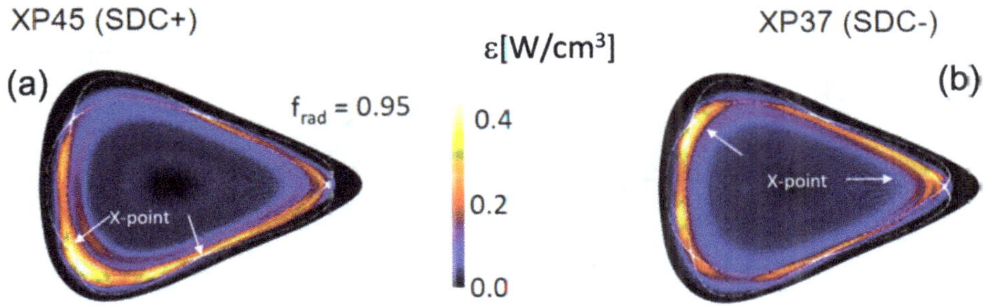

Figure 5.13. Two-dimensional radiation intensity distributions obtained by bolometer tomography in a W7-X discharge with (a) SDC+, and (b) SDC-. The white lines show the separatrix from the ideal vacuum field. Reproduced from [28].

system. Conventional three-dimensional edge transport modeling cannot reproduce the edge-localized 2D radiation pattern. The up–down asymmetry of the impurity radiation generally became more pronounced with increment of the line-averaged electron density.

As the up–down asymmetry of impurity density has been observed in magnetically confined toroidal plasmas since the early phases of research, theoretical studies on this asymmetry, i.e., on the frictional force between main and impurity ions, which is considered as the main theoretical foundation of such asymmetry, are also in progress. For tokamak plasmas, starting from the PS regime [29], neoclassical parallel transport with friction between main and impurity ions in all three collisionality regions up to the plateau [30] and banana regimes [31, 32] has been studied. (For a detailed explanation, please refer to the respective references.) For stellarator plasmas requiring more complex treatment, theoretical and numerical studies (e.g., in the PS regime [33] and so on) have also made progress. Although comparisons with experimental results have been made to verify the theoretical treatment, quantitative agreement has rarely been obtained. This is because experimental results include various effects contributing to the neoclassical parallel transport, and the impact of turbulent transport. Another reason is the lack of experimental data necessary for verification, e.g., the density and temperature of impurity ions. A complete understanding of impurity transport in magnetically confined toroidal plasmas is essential to finding effective controls for impurities in fusion reactors. Therefore, further research on poloidal variation in impurity density as it relates to transport parallel to the magnetic field is desirable.

5.4 Impurity holes

5.4.1 Discovery of the impurity hole

An impurity hole is characterized as an extremely hollow impurity density profile, observed in helical plasmas with an ITB [34, 35]. The discovery of impurity hole plasmas is important because it gave a new physical insight into the mechanisms of impurity accumulation commonly observed in tokamak plasmas with ITBs. Before the discovery of the impurity hole, the accumulation of impurities in ITB plasmas in

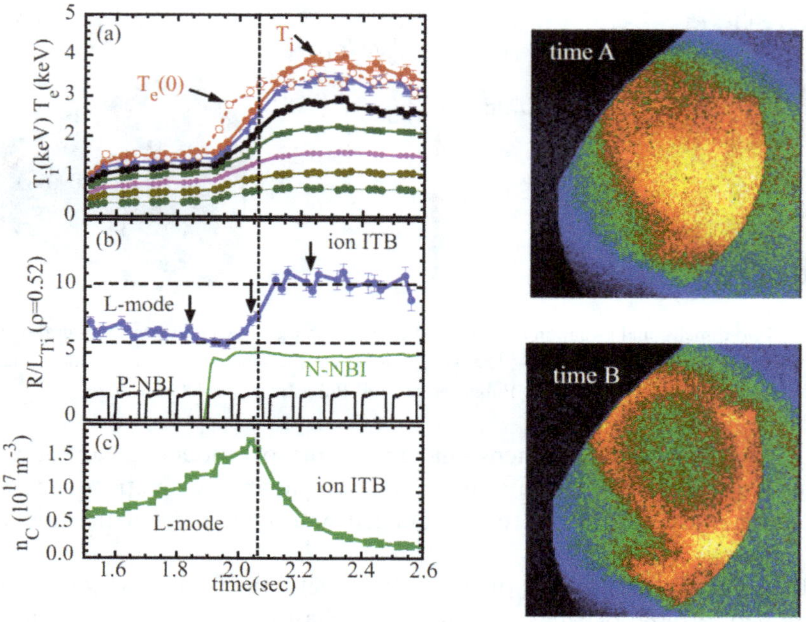

Figure 5.14. Time evolution of (a) central electron temperature and ion temperature at various radii, (b) normalized ion temperature gradient, waveform of NBI, and (c) central carbon density, as well as SXR images before (time A) and after (time B) the formation of the impurity hole. Reprinted from [34], with the permission of AIP Publishing.

tokamak devices was explained by the reduction of anomalous diffusion and the increase of inward neoclassical convection velocity due to the sharp pressure gradient inside the ITB region. In contrast, the neoclassical convection in a helical plasma with an impurity hole is inward, and the reduction of anomalous diffusion coefficient fails to explain the formation of the hole. Therefore, observation of impurity hole plasma suggests the importance of turbulence-driven impurities driven by the fluctuating electrostatic potential, as described in section 2.2.4. Figure 5.14 shows an example impurity hole observed in LHD helical plasma. The ion ITB is characterized by an increase of the peaked ion temperature due to the reduction of ion thermal diffusivity. In this discharge, a perpendicular neutral beam with a positive ion source (P-NBI) was injected and a parallel neutral beam with a negative ion source (N-NBI) was added to produce the ion ITB. The normalized ion temperature gradient (R/L_{Ti}, the ratio of major radius R to the scale length of the ion temperature gradient L_{Ti}) is a good measure for the formation of an ITB. As seen in figure 5.14(b), the jump in the normalized ion temperature gradient (R/L_{Ti}) indicates the timing of ITB formation. Prior to the formation of the ITB, the central carbon density increased constantly. However, the central C density suddenly began to decrease, associated with the formation of an ITB (t = 2.06 s), as seen in figure 5.14(c). Subsequently, the radial profile of the low-Z impurity (C) density becomes hollow.

The SXR imaging of the tangential view of the LHD clearly shows the formation of an impurity hole. The energy range of the SXR imaging system is 2.0–8.0 keV. Because the main contribution of the SXR in this energy range is the K_{α} line emissions of titanium (Ti, 4.63–4.83 keV), chromium (Cr, 5.58–5.78 keV), and iron (Fe, 6.48–6.78 keV), this tangential SXR image is a good monitor of the high-Z impurity profiles. The SXR emission is peaked at the plasma center before the formation of the impurity hole (time A). The SXR emission at time slice B, though, is extremely hollow, which indicates that the high-Z impurities which contribute to SXR emission also acquire hollow profiles.

5.4.2 Comparison of impurity transport between tokamak and helical internal transport barriers

Impurity holes are a unique feature of helical ITB plasmas. These holes are usually observed in ion ITBs in helical plasmas, but not in tokamak plasmas. To this end, impurity transport analysis for an ion ITB in both a helical plasma and a tokamak plasma was performed for a carbon impurity using charge-exchange spectroscopy, as seen in figure 5.15. The carbon impurity density becomes extremely hollow, while the electron density profile remains flat after the transition from L-mode plasma to

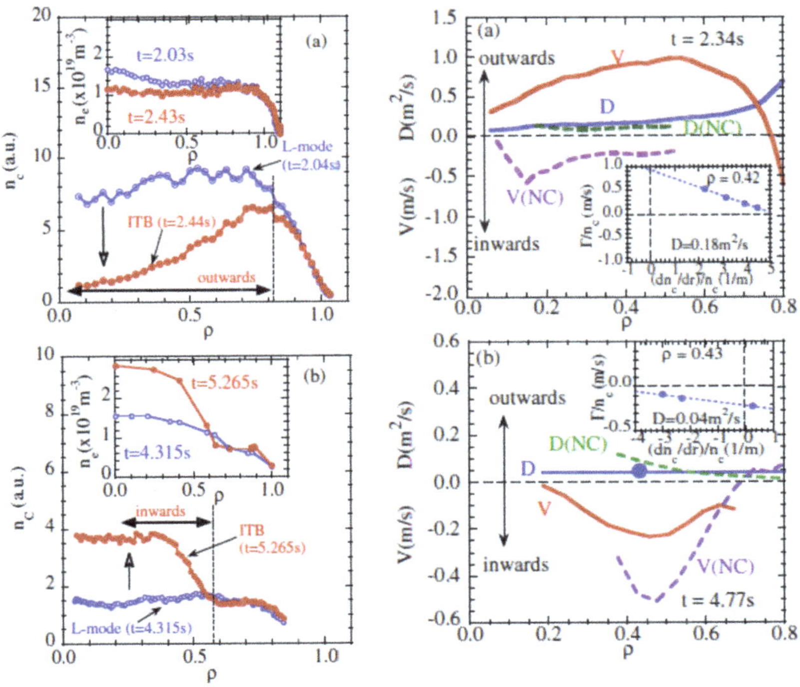

Figure 5.15. Radial profiles of electron density and carbon impurity density in (a) helical plasma, and (b) tokamak plasma, as well as radial profiles of diffusion coefficient and convection velocity in measured and neoclassical predictions. Reproduced from [36].

ITB plasma, as seen in figure 5.15(a). However, both the electron density and carbon impurity density become peaked in the tokamak ITB plasma, as seen in figure 5.15(b).

The radial flux normalized density can be expressed using the diffusion coefficient, D, normalized density gradient, and convection velocity, V_{conv}, as

$$\frac{\Gamma_z(r)}{n_z} = -D\frac{\partial n_z}{n_z \partial r} - V_{\mathrm{conv}}. \tag{5.8}$$

Therefore, the diffusion coefficient and convection velocity can be directly obtained from the flux gradient relation by fitting a data set of the normalized radial flux, (Γ_z/n_z), and normalized density gradient, $(1/n_z)(\partial n_z/\partial r)$, to a linear line, as seen in the insets in figures 5.15(c) and (d). The slope gives the diffusion coefficient, while the y-value at $x = 0$ gives the convection velocity. The diffusion coefficient derived from the slope is 0.18 $\mathrm{m^2\ s^{-1}}$ at $\rho = 0.42$ in helical ITB plasma, while it is 0.04 $\mathrm{m^2\ s^{-1}}$ at $\rho = 0.43$ in tokamak ITB plasma. These low diffusion coefficient values measured are comparable to the neoclassical diffusion coefficients.

The convection velocity evaluated is positive (outward convection) in the helical ITB plasma, while it is negative (inward convection) in the tokamak plasma. The convection velocity measured in the tokamak ITB is comparable to that predicted by neoclassical theory (at least the same sign), which is consistent with the JET results, where the time evolution of the measured impurity density profiles are consistent with those predicted by the combination of low anomalous and neoclassical convection velocity. The convection velocity measured in the helical ITB has a sign opposite to that predicted by neoclassical theory. The measured convection velocity is outward in most of the region of the plasma ($\rho < 0.75$), while the neoclassical convection velocity is inward.

5.4.3 Sign flip of convection velocity

In the steady state of an impurity hole, the outward impurity flux due to the outward convection velocity is balanced with the inward flux due to the positive impurity density gradient in the hollow impurity density profile. The sign flip of convection velocity at the beginning of impurity hole formation is abrupt and cannot be explained by the temperature screening effect as there is only a small difference in the temperature gradient before and after the formation of the impurity hole. Although several proposals have been put forward to explain the mechanism for the outward convection velocity, the physics has not been clarified yet.

Figure 5.16(a) shows the flux gradient relation of carbon impurity transport and the relation between normalized radial flux and ion temperature measured with charge-exchange spectroscopy. Before the formation of the impurity hole, both the normalized carbon density flux and normalized density gradient are negative (peaked impurity density profile). After the formation of the impurity hole, the normalized carbon density flux becomes positive. Then, the normalized density gradient subsequently becomes positive. This is important evidence showing that the normalized carbon density flux flips sign in a very short time period (0.1 s).

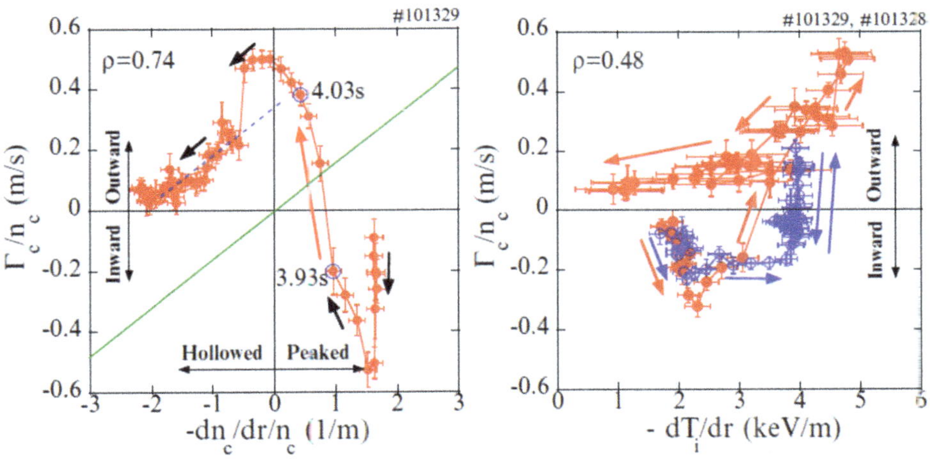

Figure 5.16. (a) Flux gradient relation of impurity transport (normalized carbon density flux vs normalized carbon density gradient), and (b) relation between normalized radial flux and ion temperature. The red symbols represent the trajectory of the radial flux for a discharge where the ion temperature gradient exceeds the critical value of 4 keV m^{-1}, while the blue data points represent the trajectory of the radial flux for a discharge with marginal temperature gradient. Reproduced from [37].

This timescale is much shorter than the timescale of change in the ion temperature gradient.

The relation between normalized radial flux and ion temperature shows clear hysteresis, as seen in figure 5.16. The sign flip of the impurity radial flux occurs when the ion temperature gradient reaches a critical value of 4 keV m^{-1}. After the sign flip of the impurity radial flux, the impurity density profile becomes hollow due to the outward convection velocity. Once the impurity profile becomes hollow, the convection velocity remains outward even when the ion temperature gradient drops below the critical value. In contrast, when the ion temperature gradient is marginal, the outward flux disappears as soon as the ion temperature gradient decreases below the critical value. One of the candidates for the mechanism causing the outward flux is potential variation on the magnetic flux surface [38]. However, a potential-variation-driven flux is not sufficient to explain the outward convection observed in experiments.

The existence of a critical ion temperature gradient and hysteresis between the radial flux and ion temperature gradient strongly suggests the non-linearity of the process causing the sign flip of the radial flux. The most possible non-linear process is the radial flux due to fluctuating potential, where the phase relation between density fluctuation and radial velocity fluctuation determines the sign of the radial flux, which can be reversed in a short timescale (see chapter 2.2.3). This is because the phase relation between the density fluctuation and radial velocity fluctuation depends on the type of turbulence, and a change in the dominant turbulence (i.e., a transition from one type of turbulence to another type of turbulence) can occur as a result of a slight change in the plasma parameters.

References

[1] Nakamura Y, Kobayashi M, Yoshimura S, Tamura N, Yoshinuma M and Tanaka K *et al* 2014 Impurity shielding criteria for steady state hydrogen plasmas in the LHD, a heliotron-type device *Plasma Phys. Control. Fusion* **56** 075014

[2] Sudo S 2016 A review of impurity transport characteristics in the LHD *Plasma Phys. Control. Fusion* **58** 043001

[3] Ida K, Yoshinuma M, Yokoyama M, Inagaki S, Tamura N and Peterson B J *et al* 2005 Control of the radial electric field shear by modification of the magnetic field configuration in LHD *Nucl. Fusion* **45** 391–8

[4] Nakamura Y, Tamura N, Yoshinuma M, Suzuki C, Yoshimura S and Kobayashi M *et al* 2017 Strong suppression of impurity accumulation in steady-state hydrogen discharges with high power NBI heating on LHD *Nucl. Fusion* **57** 056003

[5] Burhenn R, Baldzuhn J, Brakel R, Ehmler H, Giannone L and Grigull P E *et al* 2004 Impurity transport studies in the Wendelstein 7-AS stellarator *Fusion Sci. Technol.* **46** 115–28

[6] McCormick K, Grigull P, Burhenn R, Brakel R, Ehmler H and Feng Y *et al* 2002 New advanced operational regime on the W7-AS stellarator *Phys. Rev. Lett.* **89** 015001

[7] Pütterich T, Neu R, Dux R, Whiteford A D, O'Mullane M G and Summers H P *et al* 2010 Calculation and experimental test of the cooling factor of tungsten *Nucl. Fusion* **50** 025012

[8] Reiter D, Wolf G H and Kever H 1990 Burn condition, helium particle confinement and exhaust efficiency* *Nucl. Fusion* **30** 2141

[9] Campbell D J 2001 The physics of the International Thermonuclear Experimental Reactor FEAT *Phys. Plasmas* **8** 2041–9

[10] Takenaga H, Higashijima S, Oyama N, Bruskin L G, Koide Y and Ide S *et al* 2003 Relationship between particle and heat transport in JT-60U plasmas with internal transport barrier *Nucl. Fusion* **43** 1235

[11] Nakamura Y, Tamura N, Kobayashi M, Yoshimura S, Suzuki C and Yoshinuma M *et al* 2017 A comprehensive study on impurity behavior in LHD long pulse discharges *Nucl. Mater. Energy.* **12** 124–32

[12] Beidler C D, Smith H M, Alonso A, Andreeva T, Baldzuhn J and Beurskens M N A *et al* 2021 Demonstration of reduced neoclassical energy transport in Wendelstein 7-X *Nature* **596** 221–6

[13] Langenberg A, Wegner T, Pablant N A, Marchuk O, Geiger B and Tamura N *et al* 2020 Charge-state independent anomalous transport for a wide range of different impurity species observed at Wendelstein 7-X *Phys. Plasmas* **27** 052510

[14] Smeulders P 1986 Tomography of quasi-static deformations of constant-emission surfaces of high-beta plasmas in ASDEX *Nucl. Fusion* **26** 267

[15] Odstrcil T, Pütterich T, Angioni C, Bilato R, Gude A and Odstrcil M *et al* 2017 The physics of W transport illuminated by recent progress in W density diagnostics at ASDEX Upgrade *Plasma Phys. Control. Fusion* **60** 014003

[16] Chen H, Hawkes N C, Ingesson L C, von Hellermann M, Zastrow K D and Haines M G *et al* 2000 Poloidally asymmetric distribution of impurities in Joint European Torus plasmas *Phys. Plasmas* **7** 4567–72

[17] Dux R, Peeters A G, Gude A, Kallenbach A and Neu RASDEX Upgrade Team 1999 Z dependence of the core impurity transport in ASDEX Upgrade H-mode discharges *Nucl. Fusion* **39** 1509

[18] Angioni C, Casson F J, Mantica P, Pütterich T, Valisa M and Belli E A *et al* 2015 The impact of poloidal asymmetries on tungsten transport in the core of JET H-mode plasmas *Phys. Plasmas* **22** 055902
[19] Ida K and Nakajima N 1997 Comparison of toroidal viscosity with neoclassical theory *Phys. Plasmas* **4** 310–4
[20] Ingesson L C, Chen H, Helander P and Mantsinen M J 2000 Comparison of basis functions in soft X-ray tomography and observation of poloidal asymmetries in impurity density *Plasma Phys. Control. Fusion* **42** 161
[21] Reinke M L, Hutchinson I H, Rice J E, Howard N T, Bader A and Wukitch S *et al* 2012 Poloidal variation of high-Z impurity density due to hydrogen minority ion cyclotron resonance heating on Alcator C-Mod *Plasma Phys. Control. Fusion* **54** 045004
[22] Casson F J, Angioni C, Belli E A, Bilato R, Mantica P and Odstrcil T *et al* 2014 Theoretical description of heavy impurity transport and its application to the modelling of tungsten in JET and ASDEX Upgrade *Plasma Phys. Control. Fusion* **57** 014031
[23] Mynick H E 1984 Calculation of the poloidal ambipolar field in a stellarator and its effect on transport *Phys. Fluids* **27** 2086–92
[24] Ho D D M and Kulsrud R M 1987 Neoclassical transport in stellarators *Phys. Fluids* **30** 442–61
[25] Pedrosa M A, Alonso J A, García-Regaña J M, Hidalgo C, Velasco J L and Calvo I *et al* 2015 Electrostatic potential variations along flux surfaces in stellarators *Nucl. Fusion* **55** 052001
[26] Buller S, Smith H M and Mollén A 2021 Recent progress on neoclassical impurity transport in stellarators with implications for a stellarator reactor *Plasma Phys. Control. Fusion* **63** 054003
[27] Brau K, Suckewer S and Wong S K 1983 Vertical poloidal asymmetries of low-Z element radiation in the PDX tokamak *Nucl. Fusion* **23** 1657
[28] Zhang D, Burhenn R, Beidler C D, Feng Y, Thomsen H and Brandt C *et al* 2021 Bolometer tomography on Wendelstein 7-X for study of radiation asymmetry *Nucl. Fusion* **61** 115043
[29] Fülöp T and Helander P 2001 Nonlinear neoclassical transport in toroidal edge plasmas *Phys. Plasmas* **8** 3305–13
[30] Landreman M, Fülöp T and Guszejnov D 2011 Impurity flows and plateau-regime poloidal density variation in a tokamak pedestal *Phys. Plasmas* **18** 092507
[31] Helander P 1998 Bifurcated neoclassical particle transport *Phys. Plasmas* **5** 3999–4004
[32] Bielajew R and Catto P J 2023 Poloidal impurity asymmetries, flow and neoclassical transport in pedestals in the plateau and banana regimes *J. Plasma Phys.* **89** 905890411
[33] Braun S and Helander P 2010 Pfirsch–Schlüter impurity transport in stellarators *Phys. Plasmas* **17** 072514
[34] Ida K, Yoshinuma M, Osakabe M, Nagaoka K, Yokoyama M and Funaba H *et al* 2009 Observation of an impurity hole in a plasma with an ion internal transport barrier in the Large Helical Device *Phys. Plasmas* **16** 056111
[35] Yoshinuma M, Ida K, Yokoyama M, Osakabe M, Nagaoka K and Morita S *et al* 2009 Observation of an impurity hole in the Large Helical Device *Nucl. Fusion* **49** 062002
[36] Ida K, Sakamoto Y, Yoshinuma M, Takenaga H, Nagaoka K and Hayashi N *et al* 2009 Dynamics of ion internal transport barrier in LHD heliotron and JT-60U tokamak plasmas *Nucl. Fusion* **49** 095024

[37] Yoshinuma M, Ida K, Nagaoka K, Osakabe M, Takahashi H and Nakano H *et al* 2015 Abrupt reversal of convective flow of carbon impurity during impurity-hole formation on the LHD *Nucl. Fusion* **55** 083017

[38] Calvo I, Velasco J L, Parra F I, Alonso J A and Garcia-Regana J M 2018 Electrostatic potential variations on stellarator magnetic surfaces in low collisionality regimes *J. Plasma Phys.* **84** 905840407

Chapter 6

Impurity transport in the edge/scrape-off layer region

Because the magnetic field line is open in the edge/scrape-off layer (SOL) region, the impurity transport in these regions is mainly governed by the transport parallel to the magnetic field. The erosion of plasma-facing components due to physical sputtering is the main source of impurities. Because the erosion of low-Z plasma-facing materials is considerable and reduces the lifetime of the plasma wall, high-Z plasma-facing materials are considered better choices for future fusion devices. High-Z plasma-facing materials such as tungsten also have the advantage of low tritium retention limits. However, once a high-Z impurity penetrates into the last closed-flux surface (LCFS), it can cause serious problems as the impurities accumulate and radiation increases. Thus, strong suppression of this erosion by high-Z plasma-facing materials and impurity shielding in the edge/SOL region are crucial concerns.

6.1 Impurity sources in plasma-facing components

6.1.1 Physical sputtering

Tungsten (W) is a possible candidate plasma-facing material for future fusion devices because of its low tritium retention and higher threshold energy for sputtering. Figure 6.1(a) shows the sputtering yields of both beryllium (Be) and W by bulk ions (hydrogen, deuterium, and tritium) as a function of impact energy, which is strongly dependent on the plasma conditions such as the low-confinement mode (L-mode), edge-localized modes (ELMs), including inter-ELMs and intra-ELMs, hybrid high-confinement mode (H-mode scenario), etc. [1, 2]. In the L-mode, the ion energy is low enough to expect no physical sputtering of W, while the physical sputtering of low-Z plasma-facing materials such as Be show a sharp rise. Beryllium's sputtering yield peaks at ~100 eV, which correspond to the plasma conditions for the inter-ELM and hybrid scenario. In this energy range, the

doi:10.1088/978-0-7503-1451-0ch6

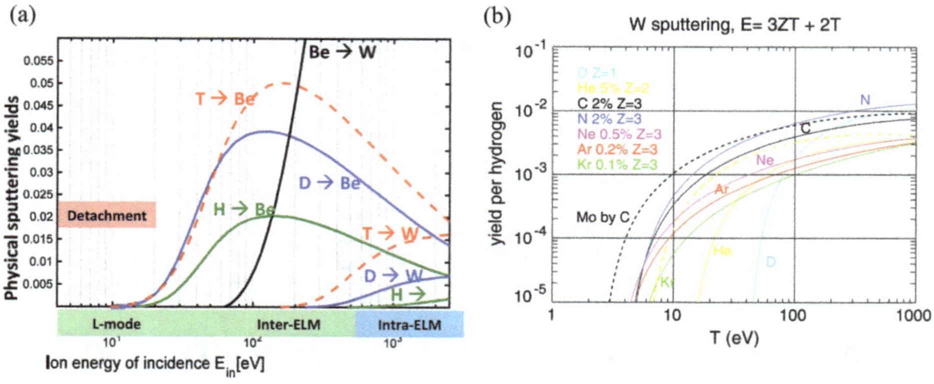

Figure 6.1. (a) Isotope (hydrogen, H: green; deuterium, D: blue; tritium, T: dotted red line) effect on beryllium (Be) and tungsten (W) sputtering yields computed by the SDTrimSP code for various phases of plasma. Note the higher W sputtering by tritium in the intra-ELM phase, as well as the increased Be sputtering by tritium, leading to an increased W sputtering by Be (in black). (b) Effective sputtering yields for W by D, helium (He), and different potential seed impurities. The molybdenum (Mo) sputtering yield by He and carbon (C) is indicated as a dotted line for comparison. Panel (a) reproduced from [1].

sputtering yields of W are negligible. They start to increase when the incident ion energy exceeds a few hundred eV, and become significant above 1 keV, which correspond to the plasma conditions for intra-ELM. The effect of isotopes on the sputtering yields is clearly observed. Those of Be by tritium (T) and by deuterium (D) are 2.5 and 2 times the sputtering yields by hydrogen (H). In contrast, the sputtering yields of W by D and T are 5–10 times that of H in the energy range for intra-ELMs. This simulation implies that the erosion of W during intra-ELMs will be a serious problem in D-T plasma mixtures in the future.

Although the high-Z material W has a higher threshold sputtering energy for bulk ion species (D), the threshold sputtering energy decreases significantly when the plasma is contaminated by impurities [3–5]. Figure 6.1(b) shows the effective sputtering yields for W by D, helium (He), and different potential seed impurities. The molybdenum sputtering yield by He and C is indicated as a dotted line for comparison. The yields have been multiplied with typical impurity concentrations expected for radiative high-performance plasmas as indicated. These effective yields per hydrogen shown give the sputtered fluxes for the seed impurity concentrations assumed if multiplied with the (hydrogenic) ion wall flux. Division of the yields shown by the indicated concentration gives the absolute sputtering yield for the impurity species. Nitrogen (N) or neon (Ne) gas seeding has been used to enhance radiation cooling at the plasma boundary or to improve performance by a combination of turbulence stabilization and improved pedestal parameters in Joint European Torus ITER like wall (JET-ILW) plasma, where high-Z plasma-facing materials are used [6]. This gas seeding causes a significant decrease in W sputtering energy by an order of magnitude. Therefore, N or Ne gas seeding enhances the W sputtering even in L-mode plasma, which would be a severe problem.

6.1.2 Chodura sheath

The larger Lamor radius of high-Z impurities reduces the net erosion of plasma-facing materials. The high-Z impurities sputtered from the plasma-facing materials are re-deposited due to a shorter ionization mean-free path length and wider Chodura sheath [9] than Debye sheath. The ionization of high-Z impurities is substantially faster than the ionization of impurities with a lower atomic number because the ionization rate of neutral impurities strongly increases with the electronic population of sputtered atoms. Therefore, the mean-free path for ionization is shorter for high-Z materials, and the reduction in net erosion of plasma-facing materials is much more significant for high-Z plasma-facing components (PFCs) than low-Z ones.

The angle between the magnetic field and the surface tangent is a key parameter to determine the impurity source at PFCs. When the magnetic field is oblique to the solid material surface, the transition layer between plasma and the material surface consists of a quasi-neutral Chodura sheath (also known as a magnetic pre-sheath) with a scale length of the order of the ion Larmor radius, ρ_i, and a Debye sheath of the scale length of several Debye lengths, λ_D, as illustrated in figure 6.2(a) [7]. The collisionless fluid model indicates that a significant electric field deflects the parallel flow in the Chodura sheath to become nearly normal at the material surface. When the sputtered material atoms are ionized within the sheath region, they can quickly be drawn back to the material surface due to the strong electric field. The magnetic field also contributes to the prompt re-deposition, especially for high-Z materials such as tungsten. Figures 6.2(b) and (c) show an illustration of the prompt re-deposition of W impurities physically sputtered from divertor targets [8]. The ionization mean-free path of neutral W and the Larmor radius of W ions are shorter than the Chodura sheath. The effective sheath electric field remains much stronger than the Lorentz force in the sheath region even for W in higher charge states.

Figure 6.3(a) shows a comparison of the potential profiles within the Chodura sheath for different α values from the solutions [10]. The potential drop across the Chodura sheath is $-kT_e/e\ \ln(\sin\alpha)$, which increases with decreasing α. The sputtered material ions with energy below the potential well of the Chodura sheath are immediately deposited to the material, which is called prompt deposition. Figure 6.3(b) shows the re-deposition rates of the W impurity, f_{redep}, of the IMPGYRO [11] results and the analytical value as a function of the W ionization length normalized by the Larmor radius of W, $\bar{\lambda}_{ionize}$. Here, $\bar{\lambda}_{ionize}$ is the ratio of the ionization length of the W neutral to the Larmor radius of W^+. The magnetic field is assumed to be a direct parallel to the surface in this analysis. The re-deposition rates of W drop to 50% within the Larmor radius of W and decrease to 10% at 3 times the Larmor radius of W. Here, the analytical value, plotted with solid lines, agrees with the IMPGYRO results, plotted with solid circles.

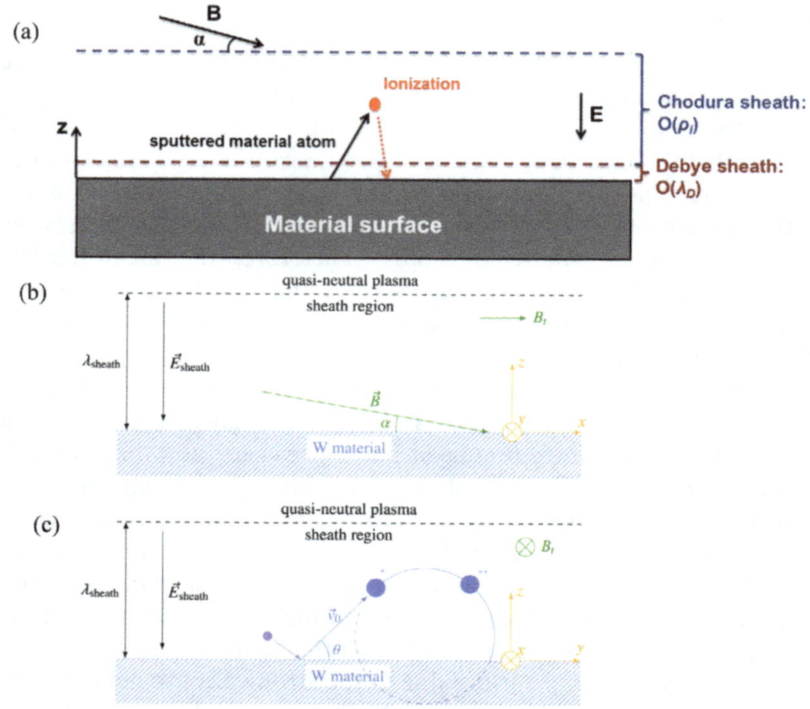

Figure 6.2. (a) Sketch of the sheath electric field near the material surface in an oblique magnetic field governing the re-deposition of sputtered particles. Illustration of the rapid re-deposition of tungsten (W) impurities physically sputtered from the target viewed from directions (b) perpendicular and (c) parallel to the oblique magnetic field. Panel (a) reproduced from [7]. © International Atomic Energy Agency. Published by IOP Publishing. All rights reserved. Panels (b) and (c) reprinted from [8], Copyright (2021), with permission from Elsevier.

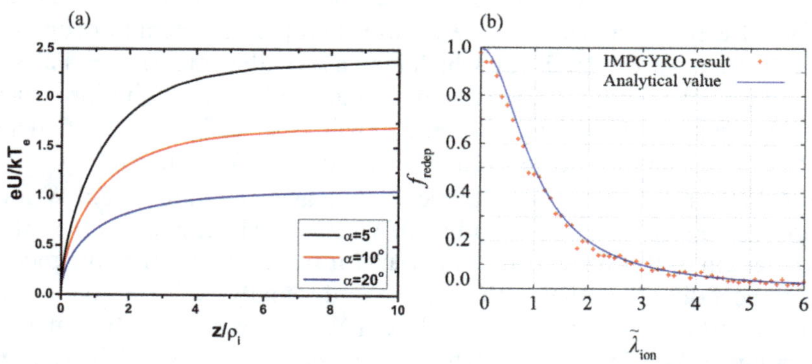

Figure 6.3. (a) Potential profiles within Chodura sheath in direction normal to the material surface for several values of α. (b) Re-deposition rates of W impurity f_{redep} of IMPGYRO results and analytical value as a function of W ionization length normalized by Larmor radius of W^+, $\tilde{\lambda}_{\mathrm{ionize}}$. Panel (a) reproduced from [7]. © International Atomic Energy Agency. Published by IOP Publishing. All rights reserved. Panel (b) reprinted from [10], Copyright (2020), with permission from Elsevier.

6.1.3 Tungsten erosion induced by edge-localized modes

ELMs are commonly observed in H-mode plasmas in tokamaks. Since the ELM causes a strong heat pulse, ELM events can erode plasma-facing materials. Several approaches exist to eliminate such ELM events and to extend the lifetime of plasma-facing materials. Resonant magnetic perturbation (RMP) has been widely used to mitigate ELM erosion. By applying RMP, tungsten erosion was demonstrated to be significantly reduced in the Experimental Advanced Superconducting Tokamak (EAST) [12], as seen in figures 6.4(a)–(e). As the current of the $n = 2$ RMP coil increases, the ELM frequency also increases from 50 Hz to 400 Hz, until, finally, the ELM disappears at $t = 5.5$ s. The peak W I emission intensity and peak probe particle flux, j_s, significantly decrease as the ELM frequency increases. This experiment shows that RMP is a valuable tool for reducing the heat load and W erosion at the divertor plate. The mitigation is seen to be more effective at the upper outer (UO) divertor.

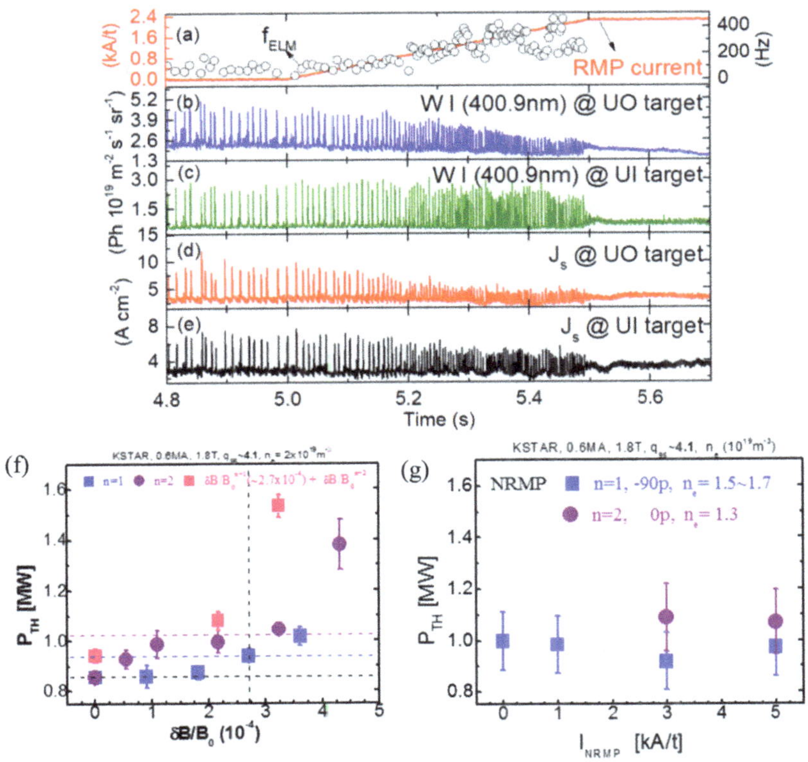

Figure 6.4. Time evolution of (a) RMP current and ELM frequency, (b) and (c) W I emission intensity, and (d) and (e) peak probe particle flux j_s at the upper outer (UO) and upper inner (UI) divertors in EAST. Threshold heating power for L- to H-mode transition as a function of fraction of perturbation field of (f) RMP and (g) NRMP. $\delta B/B_0$, in KSTAR. Panels (a)–(e) reproduced from [12]. © International Atomic Energy Agency. Published by IOP Publishing. All rights reserved. Panels (f) and (g) reproduced from [13]. © International Atomic Energy Agency. Published by IOP Publishing. All rights reserved.

However, Korea Superconducting Tokamak Advanced Research (KSTAR) experiments show that RMP increases the threshold power for the transition from L-mode to H-mode plasma, which would pose a severe problem in ITER, as seen in figure 6.4(f). When the RMP field, $\delta B/B_0$, exceeds the level of $\sim 3 \times 10^{-4}$, the H-mode threshold power increases by up to 50% for both $n = 2$ and $n = 1$ RMP [13]. These experimental results suggest that the heating power in ITER may not be enough to make a transition from L-mode to H-mode if RMP is applied for ELM mitigation. In contrast, there is no increase of H-mode threshold power observed in the case of non-resonant magnetic perturbation (NRMP) in KSTAR, as seen in figure 6.4(g). This experimental result suggests that ITER H-mode operation may not be a concern with non-resonant error field components. However, it has not been clarified whether the NRMP suppresses the W erosion because the plasma-facing material used in KSTAR is carbon. Thus, W erosion by ELM crashes is still a severe problem that has not yet been solved yet. Therefore, evaluating the magnitude of the erosion and prompt deposition in the discharge with ELM events is essential.

As described in section 6.1.1, in the intra-ELM regime, the impact (impinging) energy of incident ions exceeds the threshold energy even when using a high-Z material like W. Figure 6.5(a) shows a time evolution of the line emission of W neutral in a H-mode discharge with ELM events in the DIII-D tokamak [14]. Here, the prominent W I transition (5d $^5(^6S)$ 6s $^7S_3$ 5d $^5(^6S)$ 6p $^7P_4^o$) at $\lambda = 400.9$ nm of the sputtered W atoms is used in determining gross erosion. Figure 6.5(b) shows an expanded view of the W I emission intensity for 18 individual ELM events as a function of time relative to the ELM start. A coherent averaging with 0.1 ms binning is applied to improve the signal-to-noise ratio, which is plotted in red. The W I emission intensity rises within

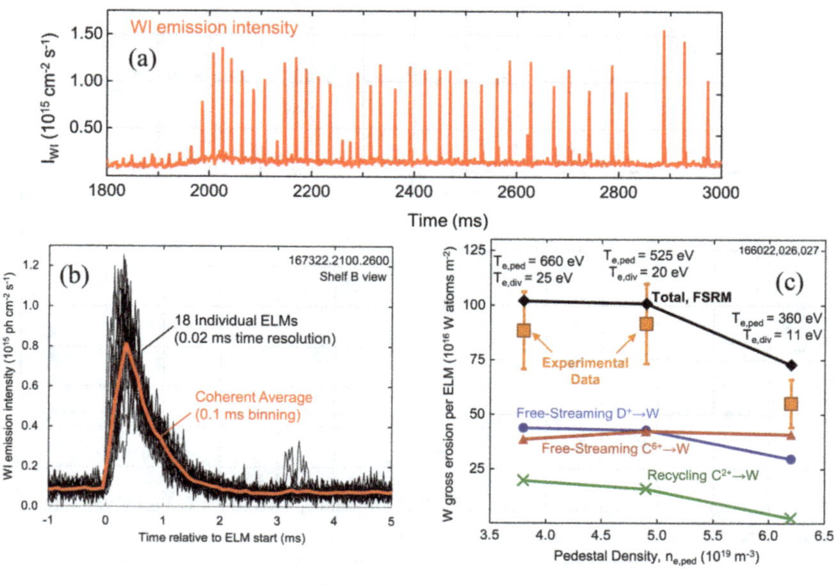

Figure 6.5. Time evolution of (a) tungsten atom W I emission intensity, (b) expanded view of W I emission intensity at ELM events, and (c) W gross erosion per ELM as a function of pedestal density. Reprinted from [14], with the permission of AIP Publishing.

0.3 ms and decays in a timescale of 1 ms. Although the duration of the W erosion is short, the peak intensity is significantly higher than that between the ELMs (inter-ELM). Figure 6.5(c) shows a comparison of the measured W gross erosion per unit area evaluated from W erosion and a prediction of the free-streaming plus recycling model (FSRM) [15]. The measured W gross erosion starts to decrease when the pedestal density exceeds 5×10^{19} m^{-3} by gas puffing. The predictions of the FSRM model, indicating the contribution of each erosion mechanism to the W sputtering rate, are overlaid. As can be seen, FSRM predicts that most of the W sputtering is caused by free-streaming D^+ and C^{6+} ions in roughly equal proportion. The contribution of the W gross erosion caused by recycling C^{2+} ions is 20% at a lower pedestal density but drops to a negligible level at a higher pedestal density of 6×10^{19} m^{-3}. The energy of the recycling D^+ ions is below the threshold for D → W physical sputtering, and, thus, no W erosion is caused by the recycling main ions. The cumulative W gross erosion calculated by the FSRM agrees with the experimental measurement.

Figure 6.6 shows a field of view of the divertor spectroscopy camera and an image of the W I emission measured with the intensified camera system [16]. Bulk tungsten and tungsten coatings on carbon-fiber-reinforced carbon (CFC) substrates are used in the area of the divertor (see also figure 6.2(a)), which is subject to high heat loads. Bulk tungsten is indicated by 'W', while the tungsten coatings on carbon-fiber-reinforced carbon substrates are indicated by 'WC'. The W sputtering intra-ELM is localized in a poloidal direction within 100 mm around the strike point at the divertor target plate, where W-coated carbon is used. It should be noted that the W sputtering from the outer strike point is much stronger than that from the inner strike point. Although the W I emission is a measure of the sputtering of W, it is not a measure of erosion due to the re-deposition of the W atoms in the Chodura sheath.

The erosion subtracting the prompt deposition and deposition during the plasma discharge from the gross erosion is called the net erosion. Because of the large Lamor radius, the net erosion is much smaller than the gross erosion for tungsten [2, 17, 18]. Figure 6.7 shows gross and net erosion profiles along the poloidal direction of the divertor target plate between the ELM (inter-ELM) and at the ELM event

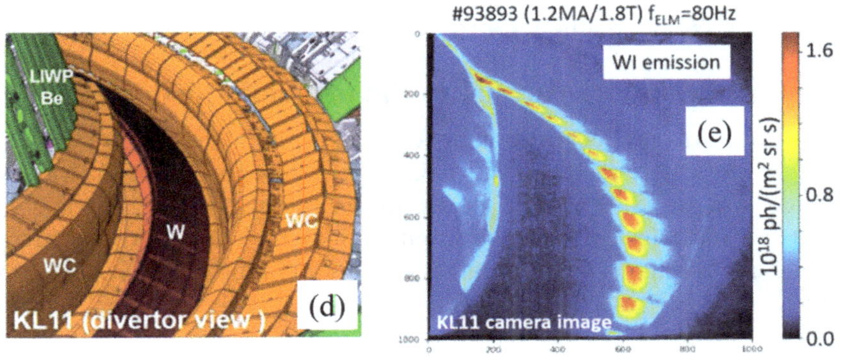

Figure 6.6. (a) Field of view of divertor spectroscopy camera, (b) single frame of intensified camera system dominated by intra-ELM tungsten (W) sputtering. Reproduced from [16].

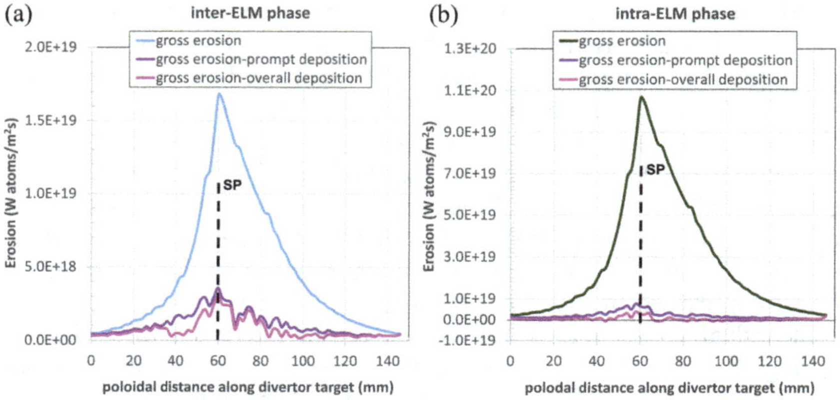

Figure 6.7. Profiles of gross erosion, gross erosion – prompt deposition, and gross erosion – overall deposition (net erosion) along the divertor tile in (a) inter-ELM phase and (b) intra-ELM phase with 50 eV and 5×10^{12} cm^{-3} at the strike point (SP). Reproduced from [17]. © Forschungszentrum Jülich GmbH. Published by IOP Publishing. All rights reserved.

(intra-ELM) with 50 eV and 5×10^{12} cm^{-3} near the strike point (indicated as SP) of the divertor target plate. The profiles indicate that the W erosion localized in a poloidal direction with the FWHM is 40 mm. The peak of the gross erosion intra-ELM is larger than that of the inter-ELM by an order of magnitude. Tungsten gross erosion is dominated in H-mode plasmas by the intra-ELM phases. The averaged (over the tile) fraction of prompt deposition is about 77% for the inter- and 93% for the intra-ELM phases. In addition to prompt deposition, further eroded particles are non-promptly deposited on the tile, resulting in an overall amount of W deposition of 84% for inter- and 98% for intra-ELM conditions. Most of the sputtered ions deposit to the materials, and the 'gross erosion – prompt deposition' and 'gross erosion – overall deposition' are much lower than the gross erosion. Due to deposition, the net W erosion can be similar within the intra- and inter-ELM phases if the inter-ELM electron temperature is high enough. This simulation shows that the deposition reduces gross erosion by a factor of 30 at the strike point under intra-ELM, and by a factor of 5 under inter-ELM conditions.

6.2 Impurity transport parallel to the magnetic field line

Impurity transport parallel to the magnetic field line in the edge/scrape-off layer (SOL) region is determined by the balance between the thermal, friction, and electrostatic forces, as described in section 2.3. Because the friction force is attributed to collisions between the impurity and bulk ions moving toward the target plate, the mechanism determining the bulk ion flow in the SOL is important for impurity transport in the SOL region.

6.2.1 Bulk ion flow in the scrape-off layer

There are several contributions to parallel flow in the SOL in a tokamak divertor configuration, which are illustrated in figure 6.8(a) [19, 20]. The ion Pfirsch–Schlüter

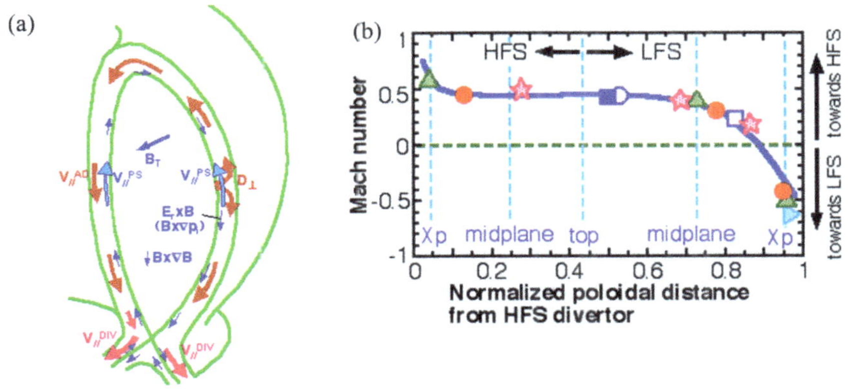

Figure 6.8. (a) Poloidal projection of parallel flow and drifts of the main ions in the SOL, and (b) typical values of the Mach number of bulk ions near the separatrix ($r^{mid} < 2$ cm) in medium $\bar{n}_e = n^{GW}$ range (0.4-0.5) for ion $B \times \nabla B$ drift direction toward the divertor. Symbols correspond to tokamaks: Alcator C-mod (star), ASDEX Upgrade (triangle), JT-60U (solid circle), Tore Supra (open circle), JET (closed square), and JET (solid square). Reprinted from [19], Copyright (2007), with permission from Elsevier.

(PS) flow, $V_{\parallel}^{PS}$, the parallel flow driven by in–out asymmetric diffusion, $V_{\parallel}^{AD}$, and the parallel flow in the divertor, $V_{\parallel}^{DIV}$, are illustrated by thick arrows. The projection of drifts in the main SOL, i.e., $E_r \times B$, $\nabla p_i \times B$, and ion $B \times \nabla B$ drift, in a normal B_t case are also shown by thin arrows. The ion PS flow is in a direction counter to the ion $B \times \nabla B$ drift and the theoretical formula in confined plasma, $V_{\parallel}^{PS} = 2q_s V_{\perp} \cos\theta$. Here, q_s is the safety factor, and θ is the poloidal angle. $E_r \times B$ is the poloidal flow in the electron diamagnetic direction, which is mainly determined by the pressure gradient of the ions. Usually, the $E_r \times B$ is parallel to the PS flow in the high-field side (HFS) and anti-parallel in the low-field side (LFS). However, the $E_r \times B$ flow is relatively small in L-mode plasma. In this case, the $B \times \nabla B$ drift is in the vertical direction toward the lower single null (LSN) divertor plate. Because the diffusion, $D_{\perp}$, has a maximum outboard midplane, the parallel flow is driven by in–out asymmetric diffusion directed toward the divertor plate, both at the HFS and LFS.

Mach probes are widely used to measure the bulk ion flow, while spectroscopic measurements of impurity lines are used to measure the impurity ion flow in the SOL. SOL flow measurements by Mach probes have been performed at different poloidal locations in many tokamaks. Figure 6.8(b) shows typical values of the Mach number of bulk ions near the separatrix ($r^{mid} < 2$ cm) in a medium electron density where the electron density, normalized by Greenwald density, $\bar{n}_e = n^{GW}$, is 0.4–0.5 in many tokamaks (e.g., JAERI Tokamak-60 Upgrade (JT-60U), Alcator C-Mod, Tokamak à configuration variable (TCV), ASDEX-Upgrade (AUG)). Here, the ion $B \times \nabla B$ drift directs toward the divertor. In the simple SOL model, the SOL flow is generated toward the divertor due to parallel gradients of plasma pressure, i.e., the parallel flow in the HFS directs downward (toward the HFS divertor), and that in the LFS also directs downward (toward the LFS divertor). However, upward SOL flow is generally observed in most regions of the LFS, except

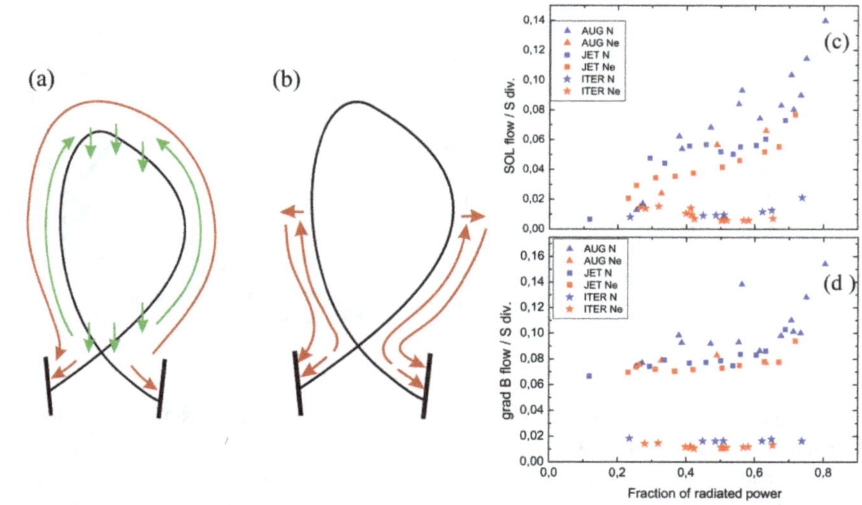

Figure 6.9. Poloidal projection of parallel SOL flow component of main ions for (a) current moderate-sized tokamaks and (b) larger tokamaks (e.g., ITER) in the future. Ratio of (c) poloidal flow through SOL and (d) $B \times \nabla B$ flow through upper main chamber separatrix to ionization in both divertor regions as functions of ratio of total radiated power to power entering computation domain with neon (Ne) and nitrogen (N) injection in ASDEX-Upgrade (AUG), JET, and ITER. Reproduced from [20]. © International Atomic Energy Agency. Published by IOP Publishing. All rights reserved.

near the LFS divertor, which is opposite to what one would expect from the simple picture. This is called flow reversal.

The contribution of parallel flow varies depending on the SOL plasma parameters, and the relative roles of various physical effects are compared between current moderate-sized tokamaks such as AUG and larger machines such as JET and ITER [20]. As seen in figures 6.9(a) and (b), the poloidal flow from the outer to inner divertors and the $B \times \nabla B$ flow contribution become smaller in a large tokamak. Here, the PS and $B \times \nabla B$ flows are shown in green, and the flows in the divertor from the ionization front toward the plates and the net flow from the outer to inner divertors in red. The drift effects and divertor asymmetries will become less pronounced for larger tokamaks in the future. The ratio of poloidal flow and $B \times \nabla B$ to the ionization source in both divertors is shown in figures 6.9(c) and (d). Two medium-Z extrinsic radiating impurity species (Ne and N) are considered. These ratios increase as the total radiated power is increased due to Ne and N seeding. Both ratios in a larger tokamak, such as ITER, are much smaller than that in current-generation tokamaks (AUG and JET). The parallel flow of the main ions directs toward the divertor target plate at the divertor leg. Therefore, the friction force between the impurity and main ions always directs toward the target plate and contributes to impurity shielding.

6.2.2 Impurity shielding by friction force

The friction force attributed to the collisions between the impurity and bulk ions moving toward the target plate increases as the ion density is increased and collision frequency increases. As aforementioned, this frictional force directs toward the target plate and contributes to impurity shielding. In contrast, the thermal force due to the temperature gradient along the magnetic field line directs toward higher temperature, i.e., upstream of the bulk ion flow. Therefore, the thermal force pushes the impurity ions to the plasma core, enhancing the impurity influx. In the fluid thermal force model, where the impurities are treated as a fluid element, the thermal force has no density dependence. The fluid mode, the ratio of frictional force, $F_{\text{frictional}}$, to thermal force, F_{thermal}, can be expressed as [21]

$$\frac{F_{\text{frictional}}}{F_{\text{thermal}}} \propto \frac{n_i|M|}{T_{\text{i}}\nabla_{\|}T_{\text{i}}}. \tag{6.1}$$

Here, M is the Mach number, and T_{i} and $\nabla_{\|}T_{\text{i}}$ are the ion temperature and ion temperature gradient along the magnetic field line, respectively. The friction force becomes dominant in higher-collisional plasma, and the thermal force becomes dominant in lower-collisional plasma. Because the friction force directs toward the divertor plate and the thermal force directs away from the divertor plate, impurity shielding can be expected in the high-collisional but not the low-collisional SOL.

The balance between thermal and friction forces determines the impurity density at the LCFS. As the electron density increases, the friction force becomes dominant, contributing to impurity shielding in the ergodic layer and the SOL. This impurity shielding has been observed in the Large Helical Device (LHD), as described in section 5.1. The disappearance of impurities even in a higher-collisionality regime, where the radial electric field is negative, described in section 5.1, can be explained by the impurity shielding due to the friction force.

The impurity shielding effect has been investigated in experiments both in tokamak and helical plasmas [22] and compared with simulations using the fluid-transport code EMC3-EIRENE [23]. In a two-dimensional (2D) distribution of force balance between thermal and friction forces, carbon density was simulated in the poloidal cross-section of the Huan-Liuqi-2A (HL-2A) tokamak for low electron density (0.25×10^{19} m^{-3}) and high electron density (0.6×10^{19} m^{-3}) at the LCFS, as seen in figure 6.10. Here, the thermal force is dominant in most of the SOL region in both cases. In the case of the lower LCFS density, the thermal force becomes more dominant in the lower half of the SOL around the main plasma. In the low-density case, thermal force dominates most of the SOL except for the vicinity of the divertor target, where the friction force is effective. Further, in the low-density case, the thermal force is significantly suppressed in the main SOL except for around the X-point, and the friction force is enhanced further below the X-point. The friction force is dominant only near the divertor target plate. The impurity shielding becomes stronger at the higher LCFS density, as seen in the significant reduction of the carbon density summed over all charge states at the SOL of the main plasma from 5×10^{17} m^{-3} to 5×10^{16} m^{-3}. The C V carbon emission line of the high-charge-state

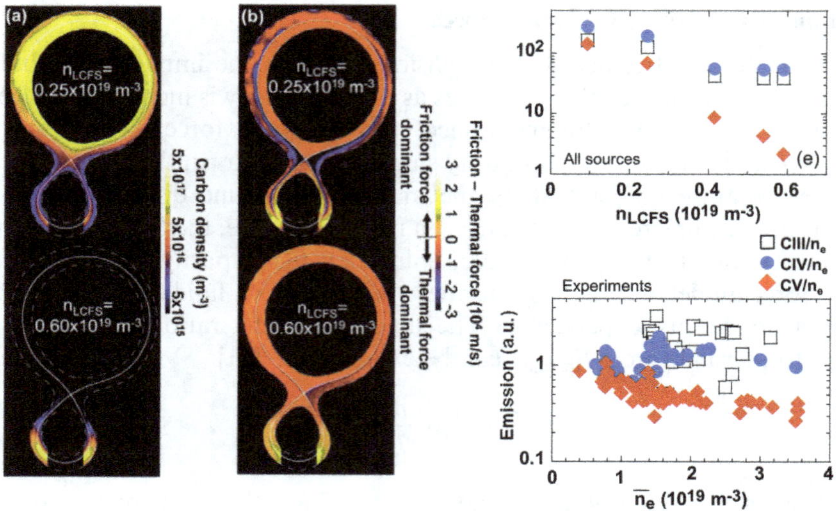

Figure 6.10. Two-dimensional distributions of (a) carbon density summed over all charge states, and (b) force balance between thermal and friction forces, as well as emissions of C III, C IV, and C V, normalized by electron density as a function of electron density, in (c) simulation and (d) experiment in HL-2A. Reproduced from [22].

carbon ions C^{4+} (with an ionization potential of 392 eV) indicates that carbon impurity penetrates into the plasma across the LCFS. In contrast, the C IV and C III carbon emission lines of the lower-charge-state carbon ions C^{3+} (65 eV) and C^{2+} (48 eV) indicate the amount of the carbon source. In the simulation, the C V emission, normalized by electron density, significantly drops (by almost two orders of magnitude) at a higher LCFS density of 0.6×10^{19} m^{-3}. In contrast, the decrease in C IV and C III emissions is moderate. This behavior is observed with spectroscopic measurements in experiment, although the impurity shielding effect is not as strong as that expected by simulation.

In a 2D distribution of the force balance between thermal and friction forces, the carbon density is simulated in the horizontally elongated poloidal cross-section of the LHD for a low electron density (2×10^{19} m^{-3}) and a high electron density (5×10^{19} m^{-3}) at the LCFS, as seen in figures 6.11(a) and (b). At the lower LCFS, the thermal force is dominant in most of the stochastic regions. At the higher density, a weakly dominant thermal force region still exists. However, a friction-force-dominant layer appears at the periphery of the stochastic layer, contributing to impurity shielding. This friction-force-dominant layer substantially impacts impurity transport parallel to the magnetic field, and reduces the carbon density from 8×10^{16} m^{-3} to 0.8×10^{16} m^{-3} by an order of magnitude. [22]. Here, the carbon density is summed over all charge states. The simulation results are compared with measurements in LHD plasma. As seen in figures 6.11(c) and (d), the normalized emission from lower-charge-state carbon ions slightly increases at higher electron density above 3×10^{19} m^{-3}. The variation of this emission is within a factor of 2 in a wide range of LCFS electron density of 1–6 $\times 10^{19}$ m^{-3}, which indicates that the

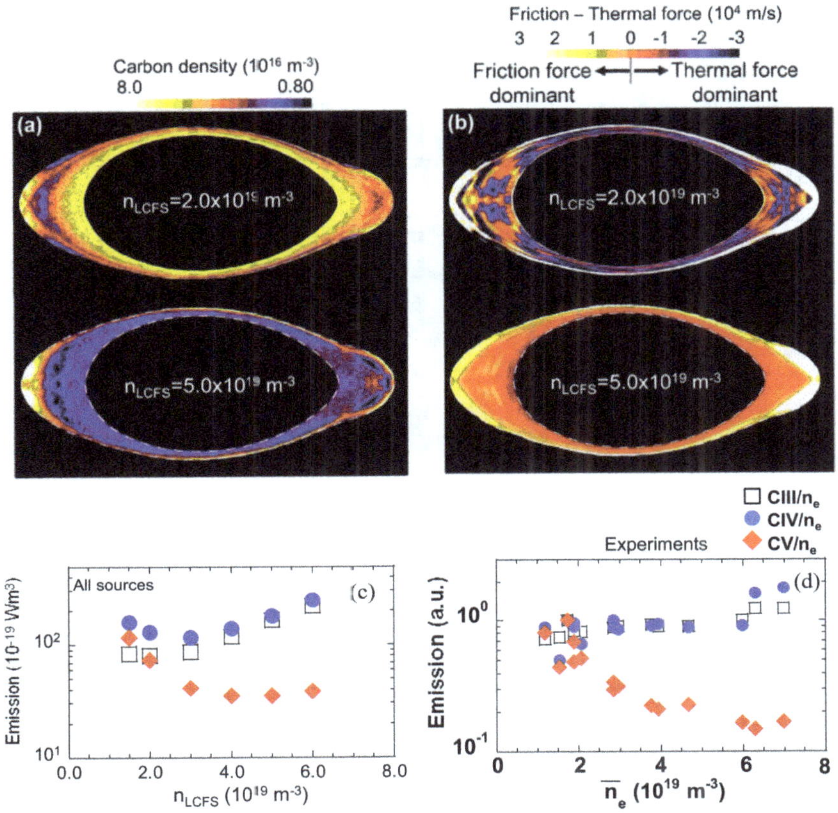

Figure 6.11. Two-dimensional distribution of (a) carbon density summed over all charge states, and (b) force balance between thermal and friction forces, as well as emissions of C III, C IV, and C V, normalized by electron density as a function of electron density, in (c) simulation and (d) experiment in LHD. Reproduced from [22].

dependence of the carbon impurity source on the LCFS electron density is relatively weak. However, the emission from the higher-charge-state carbon ions shows a substantial reduction at the higher LCFS electron density of 5×10^{19} m^{-3}. This significant drop in C V emission is due to the impurity shielding at the thin friction-force-dominant layer. The density dependence of the C V, C IV, and C III emission lines of both low- and high-charge-state carbon ions predicted by the simulation are reproduced in the experimental observations plotted in figure 6.11.

6.2.3 Impurity parallel flow: measurement and simulation

In a tokamak, the friction force dominates at the divertor leg, where the electron temperature is lower than the stochastic region. As such, the divertor leg's parallel flow is a key parameter. This parallel flow has been compared in measurement and simulation in tokamak [24–26] and helical [27, 28] plasmas. Recently, a 2D image of an impurity parallel flow was obtained using coherence image spectroscopy (CIS) at

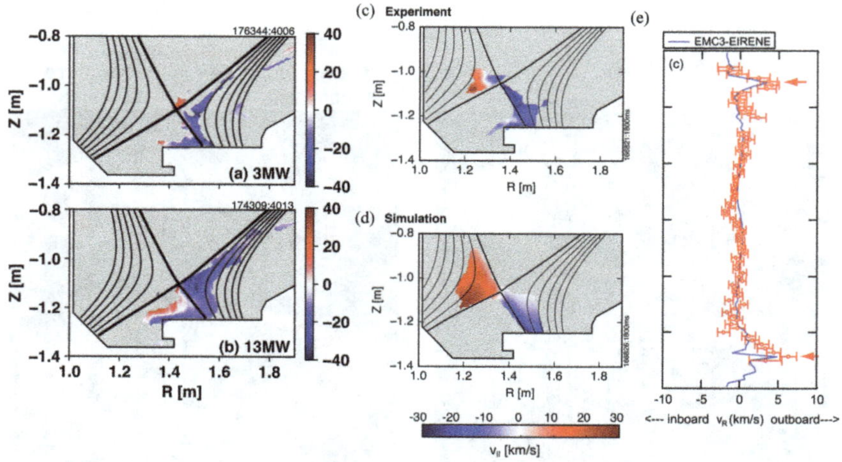

Figure 6.12. Two-dimensional image of parallel flow velocity of C^{2+} measured with C III line emission, using coherence imaging spectroscopy (CIS) in (a) 3 MW case and (b) 13 MW case in DIII-D. Two-dimensional image of parallel He^{+} velocity (c) measured experimentally with CIS, and (d) simulated using UEDGE at the lower single null (LSN) divertor configuration in DIII-D. (e) Flow velocity derived from Doppler profile of second-order C IV line emission (2 × 154.820 nm) and synthetic profile of C^{3+} flow simulated with EMC3-EIRENE code. Panels (a) and (b) reproduced from [26]. © International Atomic Energy Agency. Published by IOP Publishing. All rights reserved. Panels (c) and (d) reprinted from [25], with the permission of AIP Publishing. Panel (e) reproduced from [27]. © International Atomic Energy Agency. Published by IOP Publishing. All rights reserved.

the X-point near the lower divertor plate in the DIII-D tokamak [26]. Figures 6.12(a) and (b) show the toroidal flow of the C III impurity (C^{2+}) measured with CIS for low and high heating power. The negative flow velocity (shown in blue) is toward the divertor plate, and the positive flow velocity (in red) is away from it. The electric and friction forces overcome the thermal force, and the flow velocity toward the divertor plate is dominant, especially near the outer divertor leg. The flow velocity away from the divertor plate appears only in the private region (between the outer and inner divertor legs). The speed of the parallel flow toward the divertor plate is at the velocity of sound (Mach number ~1), and the region with significant flow velocity becomes more extensive with higher power at 13 MW. These observations demonstrate the role of impurity shielding in divertor configurations in tokamaks. This stands in contrast to the lack of impurity shielding in limiter configurations. Various optimizations of the magnetic field configuration using a magnetic island and the ergodic magnetic field produced by a perturbation magnetic field coil have been applied to prevent an impurity influx in said limiter configurations.

A direct comparison of parallel flow between measurement and simulation was done for He ions [25]. To this end, the 2D multi-fluid code UEDGE [29] is widely used. Two-dimensional parallel velocity measurements and simulations of He^{+} are given in figures 6.12(c) and (d). The positive velocity is counter-clockwise (CCW) when the machine is viewed from above. The magnetic field in the CCW direction is toward the divertor target plate in the inner leg and away from the plate in the outer leg. Then, the negative velocities in the outer leg and positive velocities in the inner leg correspond to the flow toward the divertor target plate. Velocities up to an order of $0 \sim \pm 30$ km s^{-1} are observed with negative ones in the outer leg and mostly positive ones in the inner leg. In both divertor legs, this corresponds to ion flow along field lines toward the divertor plates. Although the magnitude of the parallel flow is comparable between the inner and outer legs in the simulation, these symmetric characteristics are not observed in the experiment. One of the candidates thought to be responsible for this breaking of symmetry is the parallel flow that is coupled with the toroidal flow at the LCFS.

In helical plasma, a direct comparison of parallel flow between measurement and simulation was done for carbon ions [27] using the fluid-transport code EMC3-EIRENE. Figure 6.12(e) shows the flow velocity derived from the Doppler profile of the second-order C IV line emission (2×154.820 nm) measured by vacuum ultraviolet spectroscopy. A synthetic profile of the C^{3+} flow simulated with the EMC3-EIRENE code is also plotted with a solid line. The two solid arrows correspond to observation parallel to the SOL stochastic region at the horizontally elongated poloidal cross-section in the LHD. To obtain a synthetic vertical profile of the flow, the local value of the calculated flow projected to the line of sight is line integrated, weighted by emission intensity along each observation chord. The sign and absolute value of the plasma flow measured from the Doppler shift show excellent agreement with that predicted by the simulation code.

This good agreement of the parallel velocity between the measurements and fluid-model-based simulation indicates that both the friction and thermal forces calculated by the fluid model are valid with regards to the SOL parameters in current devices. However, as discussed in the following section, the thermal force calculation based on a fluid model would be invalid in a collisionless regime.

6.2.4 Thermal force in the low-collisional scrape-off layer

The parallel flow of impurity ions determined by the balance between the friction and thermal forces measured in experiments agrees with predictions using simulation code based on a fluid model. However, as plasma becomes less collisional the ratio of the collisional mean-free path of a focused pair of species (between bulk ions or bulk and impurity ions) to a characteristic length plasma behavior deviates gradually from the conventional fluid model description. Such deviation, called the kinetic effect due to the collisionality change, should be appropriately accommodated for in simulation models to extend the predictive accuracy to lower-collisional SOL plasmas. Recently, the kinetic effect was found to be significant enough to alter the magnitude of the thermal force, and it was pointed out that the conventional fluid model will not be sufficient to predict the parallel flow of impurity ions in a low-collisional regime where the effect is present. Therefore, the kinetic effect should be included in order to calculate the thermal force accurately, especially in future tokamak devices, where even SOL plasma is expected to become low collisional. [30–32].

Figure 6.13(a) shows the conventional and extended thermal and friction forces as a function of plasma ion density. Here, the standard fluid thermal force model is called the conventional thermal force, and the thermal force including the kinetic effect is called the extended thermal force. Fusion devices are becoming bigger and more powerful, from JT-60SA and ITER toward its successor, the DEMOnstration power plant (DEMO). Their SOL plasmas are expected to have higher temperature and lower density, as the Greenwald density limit is proportional to the inverse of the major radius of the device as $n_{\mathrm{GW}} \propto 1/R$; that is, their plasmas are expected to be less collisional. It is found that the extended thermal force is significantly reduced in the low-collisionality regime due to the kinetic effect, while the extended thermal force converges to the conventional thermal force at the collisional limit. Although the extended thermal force still overcomes the friction force, the reduction of the

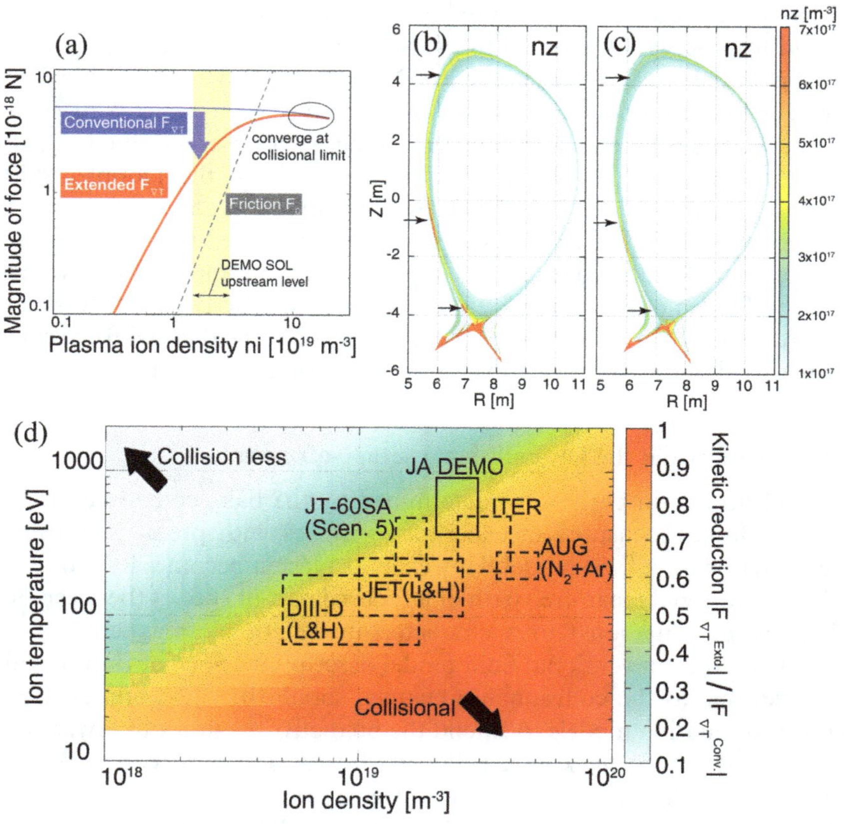

Figure 6.13. (a) Conventional and extended thermal and friction forces as a function of plasma ion density. Poloidal profiles of argon impurity density with (b) conventional thermal force and (c) extended thermal force for the DEMO fusion reactor. (d) Ratio of extended thermal force to conventional thermal force in the parameter space of ion density and temperature. Reproduced from [30], and based on figures 1, 6, 8, and 2.

thermal force is significant (20%–70%) when utilizing the SOL parameters expected for the DEMO fusion reactor.

The integrated divertor simulation code SONIC [32] was applied to study the impact on impurity transport in DEMO-like SOL plasma. SONIC consists of the fluid plasma solver SOLDOR [33], the Monte Carlo neutrals solver NEUT2D [33], and the Monte Carlo impurity solver IMPMC [34]. Figures 6.13(b) and (c) show poloidal profiles of argon impurity density with collisionality-independent thermal force (conventional thermal force) and with collisionality-dependent thermal force (extended thermal force) for the DEMO fusion reactor. An apparent decrease in impurity density was found over a wide area of the SOL upstream, especially near the X-point, midplane, and the top area in the HFS, as indicated by the three arrows in each figure. The decrease of argon density is due to the reduction of the thermal force with the kinetic effect in the low-collisional SOL in DEMO.

The ratio of the extended thermal force to conventional thermal force is also plotted in the parameter space of ion temperature and density for current tokamaks (DIII-D, JET, AUG) and next-generation devices (JT-60SA, ITER) as well as DEMO in figure 6.13(d). SOL conditions of the JET, DIII-D, and AUG tokamaks plotted in this figure are present in the discharges, where mainly an impurity-seeded H-mode could be used for ITER or DEMO-relevant studies. In the AUG tokamak, the ratio is close to unity, and the kinetic effect has a small impact on the thermal force. However, the ratio expected in JT-60SA and DEMO is in the range of 0.3–0.7, and the kinetic effect strongly impacts the thermal force. Precise modeling of the thermal force, including the kinetic effect, is a research subject of general importance for future SOL predictions of many advanced tokamak devices. JT-60SA would be an excellent device with which to test this kinetic effect in experiment.

References

[1] Joffrin E, Abduallev S, Abhangi M, Abreu P, Afanasev V and Afzal M *et al* 2019 Overview of the JET preparation for deuterium–tritium operation with the ITER like-wall *Nucl. Fusion* **59** 112021

[2] Brezinsek S, Kirschner A, Mayer M, Baron-Wiechec A, Borodkina I and Borodin D *et al* 2019 Erosion, screening, and migration of tungsten in the JET divertor *Nucl. Fusion* **59** 096035

[3] Stangeby P C and Leonard A W 2011 Obtaining reactor-relevant divertor conditions in tokamaks *Nucl. Fusion* **51** 063001

[4] Stangeby P C 2018 Basic physical processes and reduced models for plasma detachment *Plasma Phys. Control. Fusion* **60** 044022

[5] Kallenbach A, Balden M, Dux R, Eich T, Giroud C and Huber A *et al* 2011 Plasma surface interactions in impurity seeded plasmas *J. Nucl. Mater.* **415** S19–26

[6] Gabriellini S, Garzotti L, Zotta V K, Bourdelle C, Casson F J and Citrin J *et al* 2023 Neon seeding effects on two high-performance baseline plasmas on the Joint European Torus *Nucl. Fusion* **63** 086025

[7] Ding R, Stangeby P C, Rudakov D L, Elder J D, Tskhakaya D and Wampler W R *et al* 2015 Simulation of gross and net erosion of high-Z materials in the DIII-D divertor *Nucl. Fusion* **56** 016021

[8] Guterl J, Bykov I, Ding R and Snyder P 2021 On the prediction and monitoring of tungsten prompt redeposition in tokamak divertors *Nucl. Mater. Energy.* **27** 100948

[9] Stangeby P C 2012 The Chodura sheath for angles of a few degrees between the magnetic field and the surface of divertor targets and limiters *Nucl. Fusion* **52** 083012

[10] Yamoto S, Homma Y, Hoshino K, Toma M and Hatayama A 2020 IMPGYRO: The full-orbit impurity transport code for SOL/divertor and its successful application to tungsten impurities *Comput. Phys. Commun.* **248** 106979

[11] Hoshino K, Noritake M, Toma M and Hatayama A 2008 High-Z impurity transport code by Monte Carlo method in a realistic tokamak geometry – IMPGYRO – *Contrib. Plasma Phys.* **48** 280–4

[12] Chen X H, Ding F, Wang L, Sun Y W, Ding R and Brezinsek S *et al* 2021 The impact of ELM mitigation on tungsten source in the EAST divertor *Nucl. Fusion* **61** 046046

[13] Park H K, Choi M J, Hong S H, In Y, Jeon Y M and Ko J S *et al* 2019 Overview of KSTAR research progress and future plans toward ITER and K-DEMO *Nucl. Fusion* **59** 112020

[14] Abrams T, Unterberg E A, Rudakov D L, Leonard A W, Schmitz O and Shiraki D *et al* 2019 Impact of ELM control techniques on tungsten sputtering in the DIII-D divertor and extrapolations to ITER *Phys. Plasmas* **26** 062504

[15] Abrams T, Unterberg E A, McLean A G, Rudakov D L, Wampler W R and Knolker M *et al* 2018 Experimental validation of a model for particle recycling and tungsten erosion during ELMs in the DIII-D divertor *Nucl. Mater. Energy.* **17** 164–73

[16] Huber A, Brezinsek S, Huber V, Solano E R, Sergienko G and Borodkina I *et al* 2021 Understanding tungsten erosion during inter/intra-ELM periods in He-dominated JET-ILW plasmas *Phys. Scr.* **96** 124046

[17] Kirschner A, Tskhakaya D, Brezinsek S, Borodin D, Romazanov J and Ding R *et al* 2017 Modelling of plasma-wall interaction and impurity transport in fusion devices and prompt deposition of tungsten as application *Plasma Phys. Control. Fusion* **60** 014041

[18] Rudakov D L, Stangeby P C, Wampler W R, Brooks J N, Brooks N H and Elder J D *et al* 2014 Net versus gross erosion of high-Z materials in the divertor of DIII-D *Phys. Scr.* **2014** 0140

[19] Asakura N 2007 Understanding the SOL flow in L-mode plasma on divertor tokamaks, and its influence on the plasma transport *J. Nucl. Mater.* **363–365** 41–51

[20] Rozhansky V, Kaveeva E, Senichenkov I, Veselova I, Voskoboynikov S and Pitts R A *et al* 2021 Multi-machine SOLPS-ITER comparison of impurity seeded H-mode radiative divertor regimes with metal walls *Nucl. Fusion* **61** 126073

[21] Kobayashi M, Feng Y, Masuzaki S, Morisaki T, Ohyabu N and Yamada H *et al* 2008 Modelling of impurity transport in ergodic layer of LHD *Contrib. Plasma Phys.* **48** 255–9

[22] Kobayashi M, Morita S, Dong C F, Cui Z Y, Pan Y D and Gao Y D *et al* 2013 Edge impurity transport study in the stochastic layer of LHD and the scrape-off layer of HL-2A *Nucl. Fusion* **53** 033011

[23] Feng Y, Sardei F and Kisslinger J 1999 3D fluid modelling of the edge plasma by means of a Monte Carlo technique *J. Nucl. Mater.* **266-269** 812–8

[24] Petrie T W, Porter G D, Brooks N H, Fenstermacher M E, Ferron J R and Groth M *et al* 2009 Impurity behaviour under puff-and-pump radiating divertor conditions *Nucl. Fusion* **49** 065013

[25] Samuell C M, Porter G D, Meyer W H, Rognlien T D, Allen S L and Briesemeister A *et al* 2018 2D imaging of helium ion velocity in the DIII-D divertor *Phys. Plasmas* **25** 056110

[26] Petty C Cand the DIII-D Team 2019 DIII-D research towards establishing the scientific basis for future fusion reactors *Nucl. Fusion* **59** 112002

[27] Oishi T, Morita S, Dai S Y, Kobayashi M, Kawamura G and Huang X L *et al* 2017 Observation of carbon impurity flow in the edge stochastic magnetic field layer of Large Helical Device and its impact on the edge impurity control *Nucl. Fusion* **58** 016040

[28] Perseo V, Effenberg F, Gradic D, König R, Ford O P and Reimold F *et al* 2019 Direct measurements of counter-streaming flows in a low-shear stellarator magnetic island topology *Nucl. Fusion* **59** 124003

[29] Rognlien T D, Milovich J L, Rensink M E and Porter G D 1992 A fully implicit, time dependent 2-D fluid code for modeling tokamak edge plasmas *J. Nucl. Mater.* **196–198** 347–51

[30] Homma Y, Hoshino K, Yamoto S, Tokunaga S, Asakura N and Sakamoto Y *et al* 2020 Effect of collisionality dependence of thermal force on impurity transport under a lower collisional condition in DEMO scrape-off layer plasma *Nucl. Fusion* **60** 046031

[31] Reiser D, Reiter D and Tokar M Z 1998 Improved kinetic test particle model for impurity transport in tokamaks *Nucl. Fusion* **38** 165

[32] Shimizu K, Takizuka T, Ohya K, Inai K, Nakano T and Takayama A *et al* 2009 Kinetic modelling of impurity transport in detached plasma for integrated divertor simulation with SONIC (SOLDOR/NEUT2D/IMPMC/EDDY) *Nucl. Fusion* **49** 065028

[33] Kawashima H, Shimizu K and Tokamak T 2008 Analysis of particle pumping using SOLDOR/NEUT2D code in the JT-60U tokamak *Contrib. Plasma Phys.* **48** 158–63

[34] Shimizu K, Takizuka T and Kawashima H 2008 Extension of IMPMC code toward time evolution simulation *Contrib. Plasma Phys.* **48** 270–4

IOP Publishing

Impurity Transport in Magnetically Confined Plasmas

Katsumi Ida and Naoki Tamura

Chapter 7

Effect of magnetic topology on impurity transport

The effect of magnetic topology on impurity transport is described in this chapter. The impurity transport inside a magnetic island, in an ergodic or stochastic magnetic field, and in the last closed-flux surface (LCFS), the boundary between the nested magnetic flux surface and open magnetic field at the plasma periphery, are described as example topologies which have an effect on impurity transport. Good confinement of impurities is observed inside the magnetic island, while a strong impurity screening effect is identified in the ergodic or stochastic magnetic field regions. Sharp impurity density gradients are produced at the LCFS due to the boundary between two different topologies.

7.1 Magnetic island

A magnetic island is a closed magnetic flux surface bounded by a separatrix, isolating it from the rest of space. It appears at low orders of rational surface of $q = 1, 2, 3$, with poloidal (m) and toroidal (n) mode numbers of $(m, n) = (1, 1), (2, 1), (3, 1)$. Figure 7.1 shows a $m/n = 1/1$ magnetic island in plasma. The magnetic island has a crescent shape and rotates poloidally and toroidally with helical structure. The center of the magnetic island is called the O-point, while the separatrix region is called the X-point of the magnetic island. In general, the crescent shape region of the magnetic island has completely nested magnetic flux, while the magnetic field near the X-point region of the magnetic island becomes stochastic. Usually, the magnetic island is rotating in the laboratory frame due to the plasma rotation. Therefore, the signal of the magnetic field, X-ray emission, and electron cyclotron emission (ECE) from the plasma with a magnetic island oscillate, except for the case of a plasma in the locked mode, in which case the plasma rotation stops.

doi:10.1088/978-0-7503-1451-0ch7

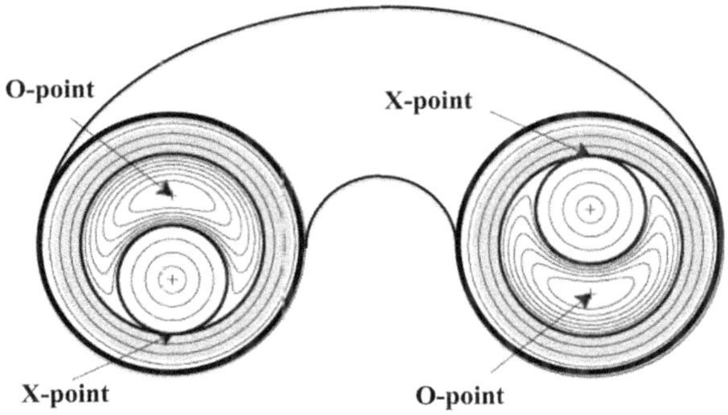

Figure 7.1. Topology of magnetic island with toroidal periodic number $n = 1$ and poloidal periodic number $m = 1$ in a tokamak device.

Usually, the electron temperature and density profiles inside a magnetic island are flat. This is not due to the poor confinement but rather to the small radial heat and particle flux inside the magnetic island. In fact, the turbulence level inside the magnetic island is even lower than the turbulence level outside the magnetic island, because there is almost no gradient to drive the turbulence. The turbulence level at the O-point of the magnetic island was found to be much lower than that at the X-point in experiments [1, 2]. A heat-pulse propagation experiment in the DIII-D tokamak demonstrated that the turbulence at the O-point of the magnetic island increases before the heat pulse reaches the O-point of the magnetic island. The increase of turbulence at the O-point and at the X-point occurs simultaneously, although the time of the heat-pulse arrival has a significant delay at the O-point due to slow heat-pulse propagation from the boundary of the magnetic island to the O-point [3]. Plasma turbulence inside the magnetic island is considered to spread turbulence from the X-point of the magnetic island not generated inside the magnetic island.

The cause of the flattening is due to the fact that the majority of the radial heat flux and particle flux are concentrated at the X-point of the magnetic island and there is only little heat flux left inside the magnetic island. The heat diffusivity inside the magnetic island is smaller than that outside the magnetic island by an order of magnitude. Several experiments have shown the good confinement (i.e., low heat and particle diffusivity) inside the magnetic island due to the low level of turbulence. The reduction of electron heat transport has been identified by the slow propagation speed of a cold pulse [4] and heat pulse [5], while the reduction of ion heat transport has been identified by the observation of a long sustained, peaked ion temperature profile inside the magnetic island after the back transition from high-confinement mode (H-mode) to low-confinement mode (L-mode), where the ion temperature at the boundary of the magnetic island suddenly drops due to the back transition. In this experiment, the effective ion thermal diffusivity inside the magnetic island was lower than that outside the magnetic island by an order of magnitude even with a finite temperature gradient inside the magnetic island similar to that outside the

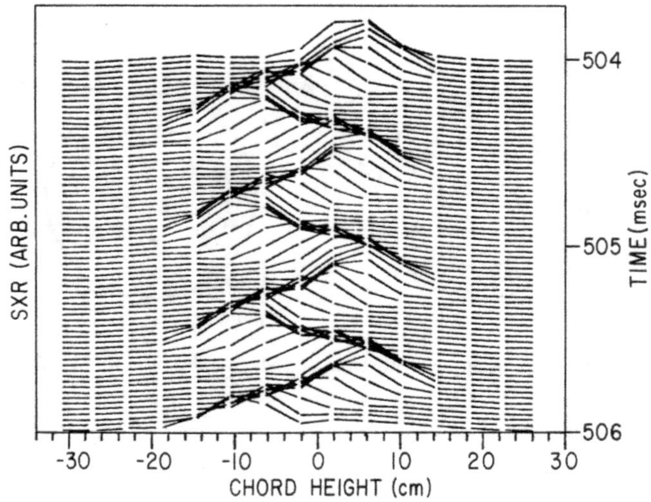

Figure 7.2. Time evolution of the radial profiles of line-integrated X-ray emission after an internal collapse event in the PBX tokamak. Reproduced from [8].

magnetic island [6]. Therefore, the good confinement inside the magnetic island is not simply due to the small gradient but also due to an effect of its topology.

Good confinement of impurities inside a magnetic island was observed in DIII-D [7] and Princeton Beta Experiment (PBX) [8] tokamak plasma as an oscillation of soft X-ray (SXR) emission localized inside the magnetic island. In DIII-D, a large $m/n = 1/1$ oscillation of central SXR emission was observed in some cases, which is known as a type-O discharge. Similar impurity behavior was observed in the PBX. Figure 7.2 shows a time evolution of the radial profiles of line-integrated X-ray emission during a large $m/n = 1/1$ oscillation ($f = 1.8$ kHz) after an internal collapse event observed in the PBX tokamak. This large $m/n = 1/1$ oscillation is observed for only a few tens of milliseconds, and a regular sawtooth oscillation starts after the stabilization of the large oscillation. While the oscillation is localized within the inversion radius, its amplitude is more or less constant between internal disruptions until it is quickly stabilized before the next internal disruption. The observed X-ray emission is contributed by the impurity line radiation rather than the continuous emission. Therefore, the oscillation represents a peaked impurity density localized inside the magnetic island and not a peaked temperature. This experiment suggested that there is good impurity particle confinement (i.e., low diffusivity) inside magnetic islands.

Because of the low level of turbulence inside the magnetic island, the particle transport is also reduced in addition to the impurity transport (i.e., a low diffusivity for the particle transport as well as impurity transport is expected). Figure 7.3 shows the pellet ablation and snake oscillation seen by the vertical SXR camera system in the Joint European Torus (JET) tokamak. The X-ray flux is shown as a function of time and detector chord radius (R) measured from the plasma center. After a hydrogen (H) pellet is injected, a snake-like perturbation is observed in the X-ray camera [9]. In this discharge, regular sawtooth oscillations are observed both before

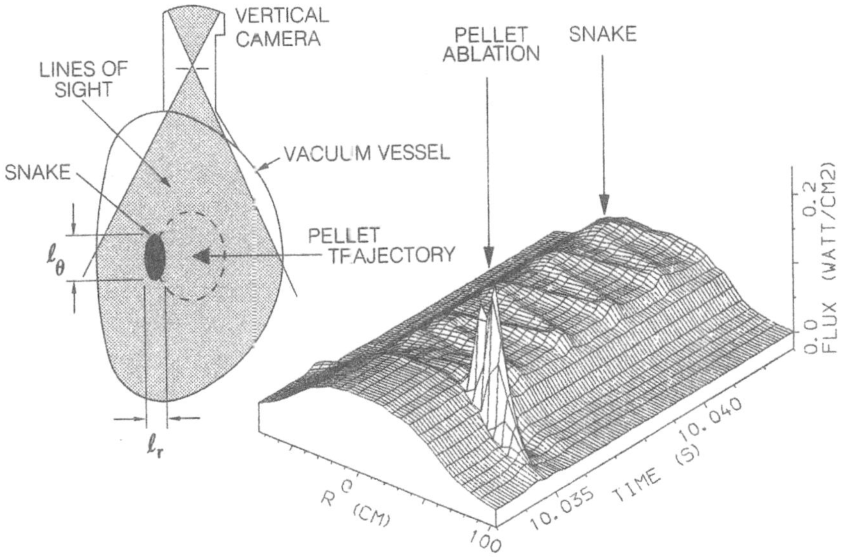

Figure 7.3. Pellet ablation and snake oscillation seen by vertical soft X-ray camera. The X-ray flux is shown as a function of time and detector chord radius (R) measured from the plasma center in JET. Reprinted with permission from [9], Copyright (1987) by the American Physical Society.

and after the pellet injection. The snake-like perturbation is due to the rotation of a small region with enhanced X-ray emission near the O-point of the magnetic island of the $q = 1$ surface. This experiment also supports a remarkably good particle confinement inside the magnetic island. It should be noted that the first observation of good confinement inside a magnetic island was identified by the oscillation of localized X-ray emission contributed by the emission from a confined impurity in the PBX tokamak. That observation came much earlier than the discovery of good particle confinement of bulk ions injected by pellet and reduced transport in heat transport, as observed in the cold-/heat-pulse propagation experiment.

An impurity injection experiment into a magnetic island in the Large Helical Device (LHD) demonstrated a longer impurity confinement time inside the magnetic island (O-point). Figures 7.4(a) and (b) show reconstructed two-dimensional computed tomography images of the absolute extreme ultraviolet photodiode (AXUVD) signal in plasma with a titanium (Ti) tracer injection at the O-point and X-point of the magnetic island. These constructed images clearly show that the Ti tracer deposition is localized inside the magnetic island for the O-point tracer-encapsulated solid pellet (TESPEL) injection, while the Ti tracer deposition is broader for the X-point TESPEL injection. The time evolution of the AXUVD signal after the TESPEL injection shows a twice-longer decay time for the O-point injection than that for the X-point injection, as seen in figure 7.4(c). The time evolution of the AXUVD signal in the case of an O-point TESPEL injection without a Ti tracer is also plotted as a reference. Good impurity confinement inside the magnetic island at the O-point is experimentally confirmed for both scenarios, using an intrinsic impurity and an externally injected impurity.

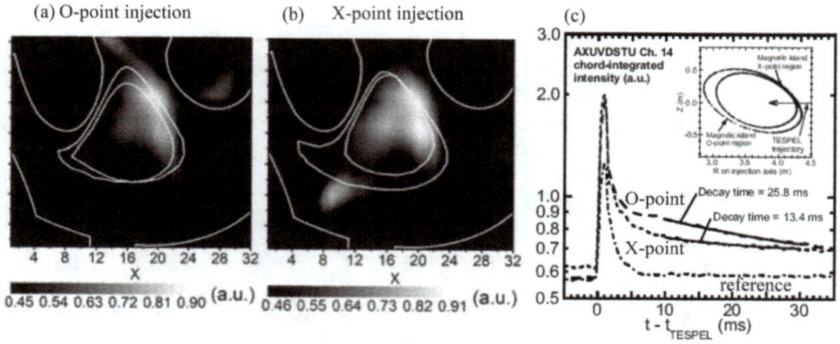

Figure 7.4. Reconstructed two-dimensional computed tomography image of an AXUVD signal in a plasma with titanium (Ti) tracer injection at (a) the O-point and (b) the X-point of the magnetic island, as well as (c) the time evolution of the AXUVD signal with respect to the time of TESPEL injection with Ti tracer. Reprinted from [10], with permission from The Japan Society of Plasma Science and Nuclear Fusion Research.

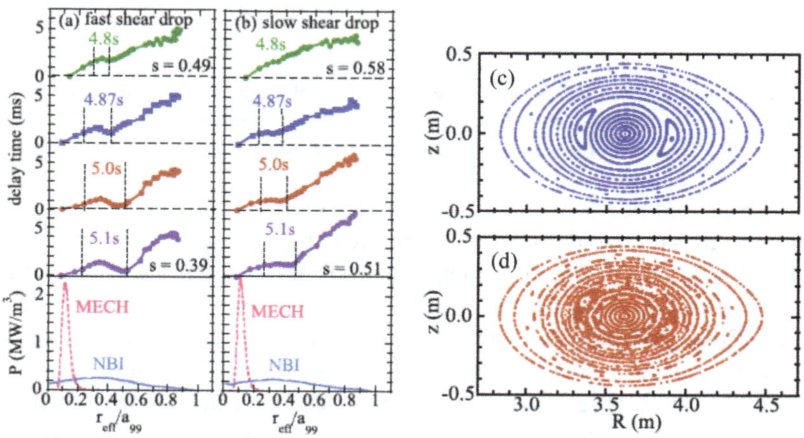

Figure 7.5. Radial profiles of the delay time of a heat pulse in plasma with (a) a nested magnetic island and (b) a stochastic magnetic island during the flattening of the electron temperature profiles, as well as Poincaré maps of the magnetic field lines calculated by 3D equilibrium code for (c) a nested magnetic island, and (d) a stochastic magnetic island. Reproduced from [11].

Although the finding of good impurity particle confinement at the O-point of a magnetic island was first achieved in the mid-1980s, good confinement inside the magnetic island was not recognized until the two states of the island (one nested and the other a stochastic magnetic field) were discovered due to the flattening of density and temperature because of a lack of radial flux across the nested magnetic flux surface. These two states were identified in a heat-pulse propagation experiment [11]. Figures 7.5(a) and (b) show radial profiles of the delay time of a heat pulse in plasma with a nested magnetic island and stochastic magnetic island during the flattening of the electron temperature profiles. Radial profiles of the deposition power density of modulation electron cyclotron heating (MECH) and neutral beam injection are also plotted. Although the flattening of the electron temperature profiles is almost identical in these two cases, the radial profile of the delay time of a heat pulse shows a clear difference. When the magnetic island is nested, the radial profile of the delay time of the heat pulse is peaked at the O-point of the magnetic island, because the heat pulse gradually propagates from the boundary of the magnetic island to the O-point of the magnetic island. However, the radial profile of the delay time of the heat pulse is flattened when the magnetic island is stochastic because of the fast heat-pulse propagation along the stochastic magnetic field lines. Figures 7.4(c) and (d) show Poincaré maps of the magnetic field lines calculated by three-dimensional (3D) equilibrium code in plasmas with a $m/n = 2/1$ magnetic island and a stochastic region with $m/n = 2/1$, 4/2, 6/3, and 8/4 perturbations of toroidal current using the iota profile consistent with the measurements. This experiment clearly shows that there is a bifurcation between the two magnetic island states: one is with a nested flux surface and the other is a stochastic magnetic field.

7.2 Edge stochastic magnetic field region

7.2.1 Ergodic divertor in tokamak plasma

In plasma with toroidal symmetry, the magnetic field provides the magnetic flux surface, where the magnetic field lines do not cross anywhere. In a toroidal geometry, these magnetic flux surfaces with different pitch angles are nested together. However, the nested magnetic flux surfaces are broken by a tiny magnetic field perpendicular to the flux surface, and a stochastic magnetic field region appears in the plasma. Even in tokamak plasmas with toroidal symmetry, this stochastic magnetic field region, the so-called ergodic layer, can be produced by the perturbation coils installed into the vacuum vessel. Because the boundary layer in tokamaks will be only 1 (or a few) cm thick in a high-magnetic-field device, the heat flux of this scrape-off layer (SOL) is highly concentrated in a narrow region. The ergodic layer near the plasma boundary can contribute to the expansion of this narrow region. Hence, the concept of an ergodic divertor (ED), where an ergodic layer is produced by the perturbation magnetic field near the plasma boundary, is introduced.

Figure 7.6(a) shows a divertor structure in light of C III and the structures obtained from Poincaré mapping in Tokamak Experiment for Technology Oriented Research (TEXTOR) [12, 13]. On the right, the dynamic ergodic divertor (DED) coils are indicated, where the two colors represent the directions of the electric current. Recycling structures are observed in the C III intensity (white) near the DED coils. The Poincaré map of the magnetic field lines is indicated in red. A scheme of the heat flows in a stochastic magnetic field (ergodic layer) is outlined in figure 7.6(b). The radial heat flux is considerably enhanced due to the parallel heat flux along the magnetic field, which has a large radial excursion. Figures 7.6(c) and (d) show a time evolution of the DED current, C III and C V emissions, and a concentration of C^{6+} with DED current. As the DED current increases during the discharge, the C III emission increases, though the C V emission decreases. Because the intensity emission of partially ionized impurity lines is sensitive to electron temperature in the plasma, the simultaneous decrease and increase in intensity of

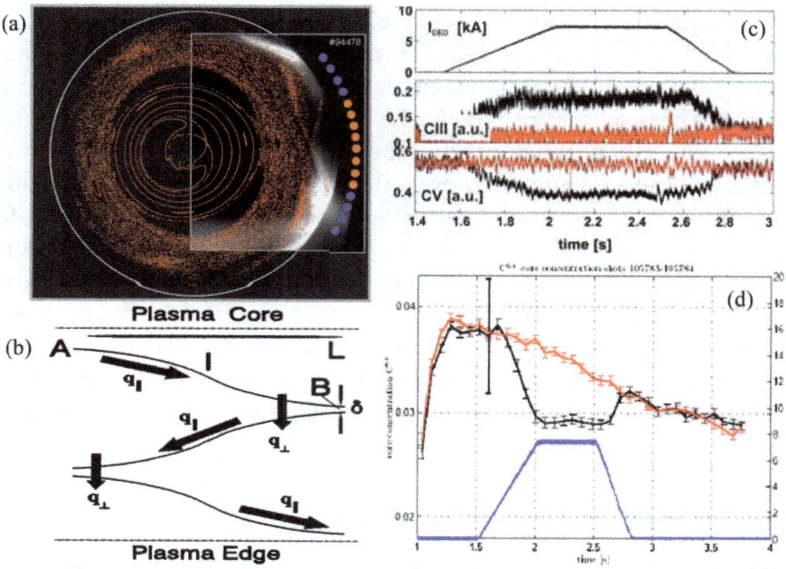

Figure 7.6. (a) Divertor structure in light of C III and the structures obtained from Poincaré mapping in TEXTOR, (b) scheme of heat flows in a stochastic magnetic field, as well as time evolution of (c) DED current, C III and C V emissions, and (d) DED current and concentration of C^{6+}. Panel (a) reproduced from [12]. Panel (b) reproduced from [13]. Panels (c) and (d) reprinted from [14],

these emissions can be explained by the change of electron temperature and/or impurity concentration. In order to study the reduction of impurity concentration (i.e., the impurity screening effect), fully ionized carbon should be investigated. The concentration of C^{6+}, which is insensitive to the temperature of the plasma, clearly decreases when a DED is applied to the impurity in the plasma core, as can be seen in the black line in figure 7.6(d). The time evolution of the concentration of C^{6+} without DED is also plotted in red as reference. This experimental result demonstrates an impurity screening effect due to the ergodic magnetic field produced by the DED.

Figure 7.7(a)(b) shows Abel-inverted emissivity profiles of C VI (3.374 nm) and C V (4.020 nm) before and during ergodic divertor (ED) perturbation and with and without ED perturbation in the Tore Supra tokamak. The decreases of emissivity due to ED perturbation is observed both in C VI and C V. Peak of the emission moves slightly inward, which imply the slight decrease of electron temperature in the plasma. The reduction of emission by a factor of three clearly demonstrates the impurity screening effect by ED perturbation. Figures 7.7(c)–(e) show the C VI emission and O VIII emission normalized by square of electron density, which is proportional to carbon and oxygen concentrations at fixed temperature, and Z_{eff} as a function of the average electron density with and without ED perturbation in the Tore Supra tokamak. The carbon concentration shows a significant reduction by ED perturbation, while the oxygen concentration actually increases by ED perturbation. The concentration of carbon and Z_{eff} decreases as the averaged density is increased.

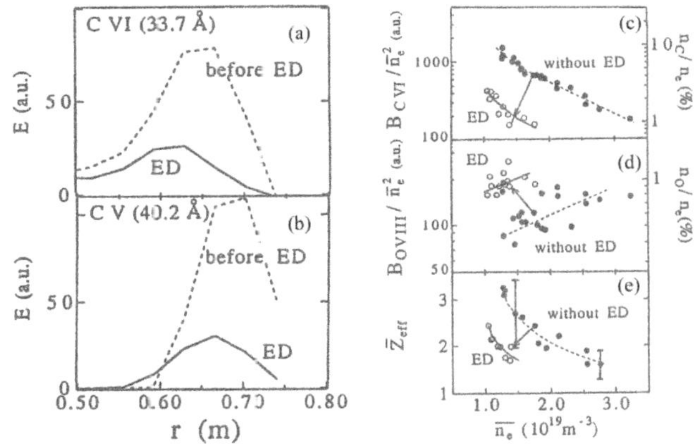

Figure 7.7. (a) Abel-inverted emissivity profiles of (a) C VI (3.374 nm) and (b) C V (4.020 nm) before and during ergodic divertor (ED) perturbation, as well as (c)–(e) the carbon and oxygen concentrations and Z_{eff} as a function of the average electron density with and without ED perturbation in Tore Supra. Reproduced from [15].

This is because the impurity screening effect increases at higher collisionality. Further decrease of Z_{eff} is observed when ED perturbation is applied, which indicates the impurity screening effect is enhanced by ED perturbation. The significant reduction of carbon density and Z_{eff} using EDs was also demonstrated in TEXTOR and Tore Supra.

7.2.2 Intrinsic stochastic magnetic field in helical plasma

The stochastic magnetic field region in the high-temperature plasma core can be identified by the heat-pulse experiment because the heat-pulse propagates radially at the speed of sound along the magnetic field, which is much faster than the heat-pulse propagation across the magnetic field line by more than one order of magnitude [11]. In the stochastic region in the plasma core, the radial profiles of electron temperature, ion temperature, and toroidal flow velocity become flat and the timescale of the heat-pulse propagation is less than a millisecond [17]. The radial electric field is almost unchanged when the stochastic magnetic field appears only in the plasma core surrounded by the nested magnetic flux surface. Figures 7.8(a)–(d) show Poincaré maps of magnetic field lines calculated by 3D equilibrium code in plasmas with a vacuum magnetic axis of 3.75 m and 3.85 m without and with a local island divertor (LID). The stochastic magnetic field regions are expanded by applying the perturbation produced with the magnetic field of the LID coil. Figures 7.8(e) and (f) show radial profiles of edge electron temperature with a vacuum magnetic axis of 3.75 m and 3.85 m. A flat region of the electron temperature profile appears with the LID. It should be noted that the temperature gradient outside the LCFS is almost unchanged in the plasma with a vacuum magnetic axis of 3.75 m, while the temperature gradient outside the LCFS decreases in the plasma with a vacuum

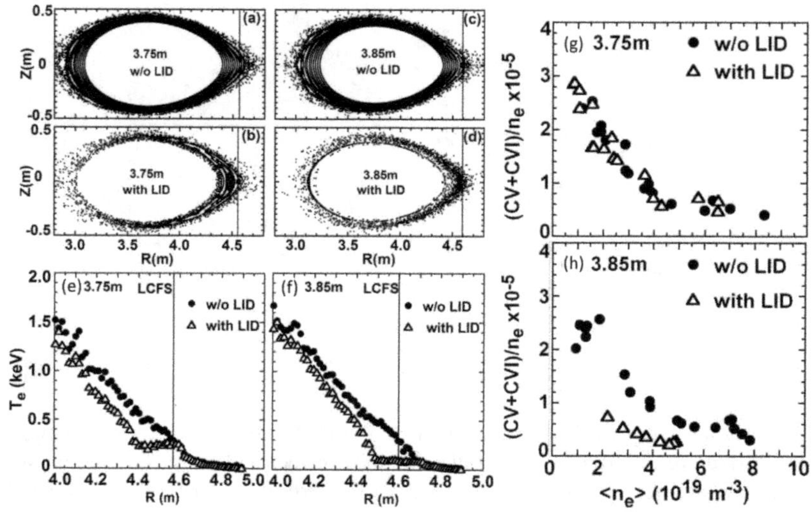

Figure 7.8. Poincaré maps of magnetic field lines calculated by 3D equilibrium code in plasmas with a vacuum magnetic axis of (a) and (b) 3.75 m and (c) and (d) 3.85 m, (a) and (c) without and (b) and (d) with local island divertor (LID), as well as radial profiles of edge electron temperature with a vacuum magnetic axis of (e) 3.75 m and (f) 3.85 m and (g) and (h) intensity of C V + C VI normalized by electron density, n_e, as a function of averaged electron density $\langle n_e \rangle$. Reprinted from [16], with the permission of AIP Publishing.

magnetic axis of 3.85 m by LID. Figures 7.8(g) and (h) show the intensity of C V + C VI normalized by electron density, n_e, as a function of averaged electron density, $\langle n_e \rangle$ [18]. The intensity of C V + C VI sharply decreases as the averaged density is increased, which is similar to the ED experiment in tokamak devices. The impact of the LID on the carbon concentration is not visible in the plasma with a vacuum magnetic axis of 3.75 m, where there is no change in temperature gradient outside the LCFS. In contrast, the LID contributed to a significant reduction of carbon density in the plasma with a vacuum magnetic axis of 3.85 m, where the temperature gradient outside the LCFS almost vanishes. Therefore, the significant reduction of carbon density of the LID is considered to be due to the disappearance of the thermal force.

In order to evaluate the concentration of impurities in the plasma more precisely, the fully ionized carbon density should be measured rather than the intensity of the partially ionized carbon density such as C^{4+} and C^{5+} from C V and C VI intensity, because the intensity of C V + C VI depends on the electron temperature as well as the content of the intensity. Although fully ionized impurity or bulk ions do not emit line radiation, their density can be measured using charge-exchange spectroscopy, as described in section 3.3. Hydrogen (H^+) and helium (He^{2+}) were measured with charge-exchange spectroscopy using a two-wavelength spectrometer [20] in the LHD. Figure 7.9 shows Poincaré maps of magnetic field lines calculated by 3D equilibrium code in plasmas as well as the radial profiles of He impurity for two different magnetic field configurations with a vacuum magnetic axis of R_{ax} = 3.6 m and 3.9 m. One is a configuration with a thin stochastic magnetic field region

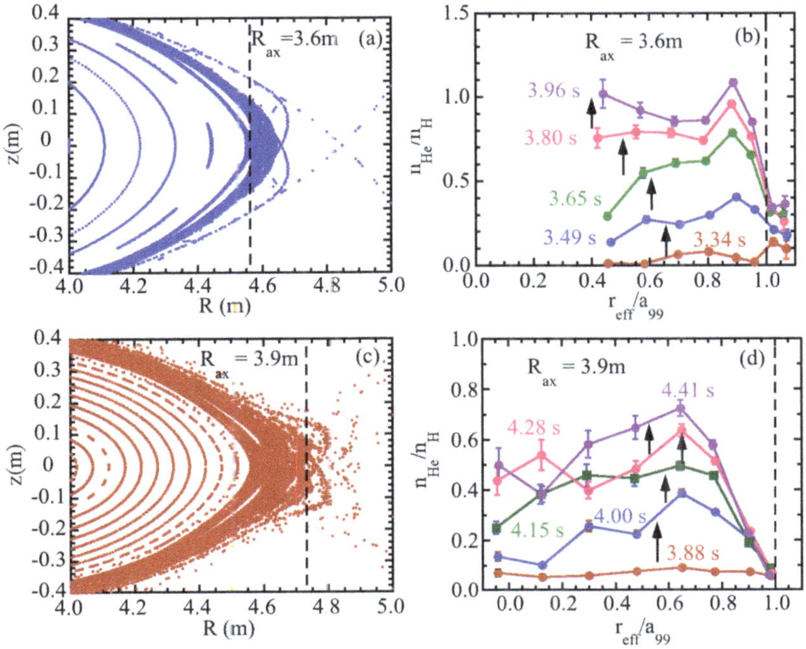

Figure 7.9. Poincaré maps of magnetic field lines calculated by 3D equilibrium code in plasmas with a vacuum magnetic axis of (a) 3.6 m and (c) 3.9 m, as well as radial profiles of helium-to-hydrogen density ratio in plasmas with a vacuum magnetic axis of (b) 3.5 m and (d) 3.9 m. Reproduced from [19].

(R_{ax} = 3.6 m), and the other is a configuration with a thick stochastic magnetic field region (R_{ax} = 3.9 m), as seen in the Poincaré maps of the magnetic field. The dotted lines show the plasma boundary of $r_{eff}/a_{99} = 1$. a_{99} is the effective minor radius, where 99% of the kinetic energy is confined. In the configuration with a magnetic axis of R_{ax} = 3.9 m, a significant stochastic region exists inside the plasma boundary of $r_{eff}/a_{99} = 1$. A He gas puff is applied to the plasmas with these magnetic field configurations to study the penetration of the impurity into the plasma core. After the He puff, the helium-to-hydrogen density ratio increases in time, as seen in the series of time slices in the radial profiles. The amount of He is larger in the configuration with a thin stochastic magnetic field region than in that with a thick stochastic magnetic field region. The He density in the stochastic magnetic field region is much lower than that in the plasma core. This experiment demonstrates that the stochastic magnetic field region very effectively prevents the penetration of impurities.

7.3 Last closed-flux surface

7.3.1 Structure of radial electric field at last closed-flux surface

In general, the radial electric field becomes positive in the SOL, where the magnetic field line connects to the limiter, divertor, or vacuum vessel wall. This is because the

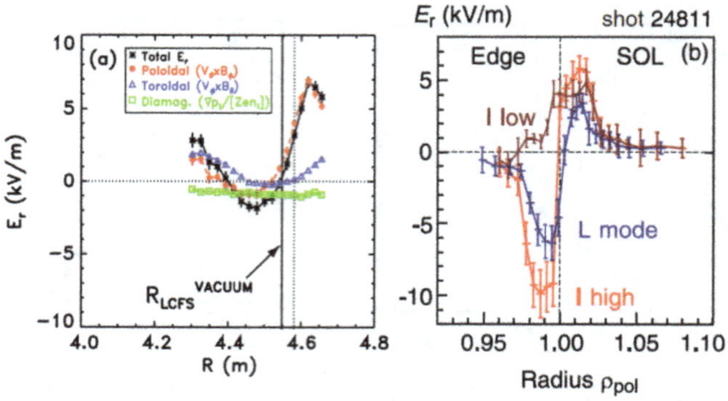

Figure 7.10. (a) Radial profile of radial electric field in plasma with stochastic plasma boundary in the LHD helical device, and (b) radial profile of radial electric field in plasma without stochastic plasma boundary in the ASDEX Upgrade tokamak. In (a), the black line represents the radial electric field, while the red, blue, and green lines represent the poloidal rotation, toroidal rotation, and diamagnetic terms of a radial electric field. Panel (a) reproduced from [21]. © International Atomic Energy Agency. Published by IOP Publishing. All rights reserved. Panel (b) reprinted with permission from [22], Copyright (2011) by the American Physical Society.

electrons escape more easily than ions along the magnetic field line, and the positive radial electric field is produced in the open magnetic field region. When the stochastic magnetic field region locates near the plasma periphery, some of the magnetic field lines are connected to the wall and the others are closed. Because the electrons escape along the open magnetic field in the stochastic region, the radial electric field also becomes positive in the stochastic magnetic field region. These positive radial electric fields contribute to exhausting the impurity, as described in section 5.1.

Figure 7.10 shows radial electric fields in LHD helical plasma where the magnetic field is stochastic near the boundary of the plasma ($R = 4.55$ m) and in Axially Symmetric Divertor Experiment (ASDEX) Upgrade tokamak plasma where there is no stochastic magnetic field at the boundary ($\rho_{\text{pol}} = 1.0$). In the helical plasma, the radial electric field becomes positive at the stochastic region ($R > 4.55$ m), and the dotted line shows the plasma boundary defined by the magnetic flux surface where 99% of the kinetic energy is confined. The radial electric field inside the LCFS depends on the plasma collisionality. In a low-collisionality regime, the radial electric field becomes positive due to the non-ambipolar electron loss, which is known as electron root. In contrast, in a high-collisionality regime, the radial electric field becomes negative, which is known as ion root. The plasmas plotted in figure 7.10(a) are in the ion-root regime, and the radial electric field is slightly negative just inside the plasma boundary ($R < 4.55$ m). The radial electric field outside the LCFS is always positive regardless of the collisionality regime (both in plasma with electron and ion roots). In tokamak plasmas, the radial electric field alters its sign from negative to positive at the boundary of the plasma as seen in figure 7.10(b). The radial electric field is strongly negative just inside the boundary, and is moderately positive outside the boundary where the magnetic field becomes open (SOL) in the L-mode. In the intermediate phase (I-phase), the radial electric field just inside the boundary of the plasma is oscillating between positive in the I-low phase and negative in the I-high phase. Because a positive radial electric field contributes to preventing the influx of impurities near the boundary of the plasma, a stochastic magnetic field is useful for reducing the concentration of impurities in the plasma core.

The parallel SOL flow driven by perpendicular transport can increase or decrease the positive radial electric field in the SOL depending on the topology of the magnetic field in the divertor configuration. Usually, the divertor target plates are located at the upper inner wall and lower inner wall in tokamak devices. There are two possible configurations in a tokamak: one is a lower single null (LSN) configuration, where the X-point is created near the lower inner divertor target plate, and the other is an upper single null (USN) configuration, where the X-point is created near the upper inner divertor target plate. Figure 7.11 shows the parallel flow driven by perpendicular transport in the SOL in LSN and USN configurations. In both configurations the SOL flow is from the midplane of the torus to the X-point [23]. As a result of transport-driven parallel flows, SOL plasmas can develop a net-volume-averaged toroidal momentum. In LSN configurations (left figure), the net-volume-averaged toroidal flow becomes parallel to the plasma current (in the co-current direction; CCW direction in the figure). In contrast, net-volume-averaged toroidal flow becomes anti-parallel to the plasma current (in the counter-current direction; CW direction in figure) in USN configurations (right figure). Because the toroidal flow in the co-current direction contributes the positive radial electric field, the positive electric field becomes stronger in LSN configurations and weaker in USN configurations. Therefore, the toroidal flow and radial electric field near the separatrix are strongly affected by the topology of the magnetic field in the SOL region. These topology-dependent SOL flows can be a source of the overall rotation drive [24] and influence the confinement of the plasma through the edge radial electric field and its shear. The magnetic topology is well known to affect the power required for the transition from an L-mode to a H-mode (otherwise known as the L–H threshold power) [25]. The L–H threshold power is lower when the ion ∇B drift (in the $B \times \nabla B$ direction) is toward the X-point (LSN configuration). Because the radial electric field is negative inside the LCFS, the parallel SOL flow driven by perpendicular transport contributes to an increase of the radial electric field shear in LSN configurations, while it contributes to a decrease of the radial electric field shear in USN configurations. Therefore, parallel SOL flow driven by perpendicular transport is one of the candidates for explaining the L–H threshold power dependence on magnetic topology (the location of the divertor configuration X-point in tokamaks).

Figure 7.12 shows the distribution of the radiated power density by argon (Ar) species in the computational region in modeling for double null (DN), LSN, and USN configurations [26]. The solid red lines indicate the divertor entrance in the simulation. The input power and the density at the low-field side midplane separatrix are the same for all three configurations. In the LSN setup, the $B \times \nabla B$ direction drift of ions is pointed to the X-point of the lower divertor. Argon radiation in the private region of the divertor increases along the divertor leg after an Ar gas puff.

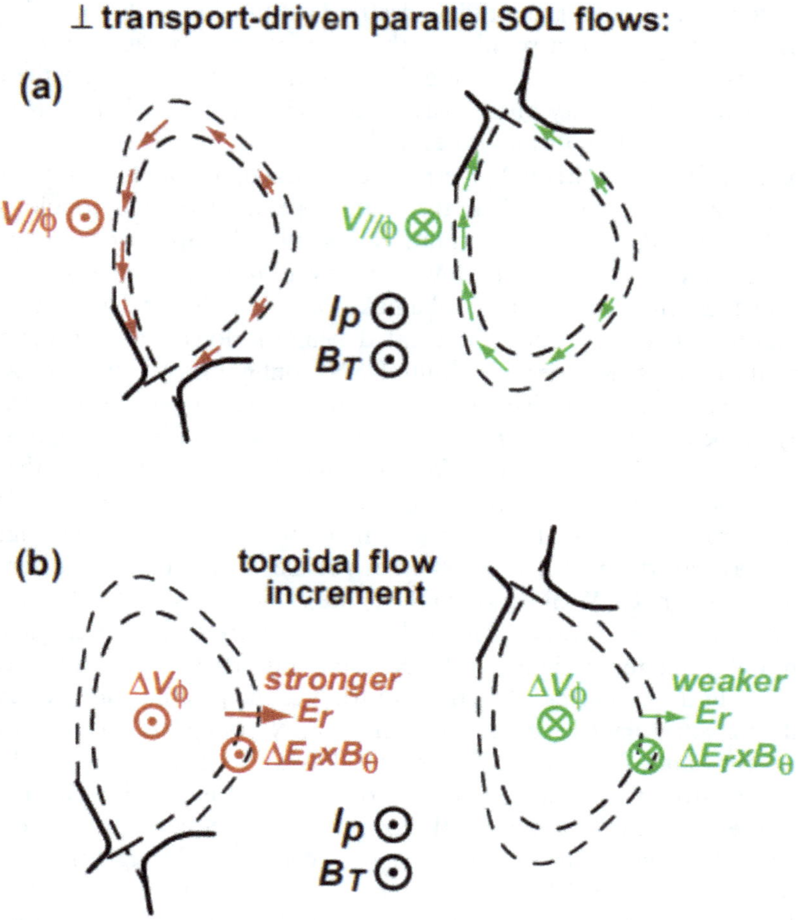

Figure 7.11. (a) Parallel flow driven by perpendicular transport in the scrape-off layer (SOL) in the lower single null and upper single null configurations, and (b) the impact of the parallel flow on the radial electric field in the SOL. Reproduced from [23].

There are significant differences in the penetration of Ar impurity in the SOL region around the plasma periphery between LSN and USN configurations. Only a small increase of Ar radiation is seen in the SOL region in the LSN configuration, but a strong increase of Ar radiation is present in the SOL region around the plasma periphery in the USN configuration. The experiment and modeling suggest that Ar seeded from the lower divertor into the plasma by the LSN magnetic configuration may have the lowest Ar penetration into the core and the strongest divertor radiation increase due to Ar seeding. These differences suggest that the Ar impurity is effectively shielded in the LSN configuration, but that the Ar impurity penetrated into the separatrix region in the USN configuration. This stronger positive electric field in the LSN configuration is one of the candidates to help prevent the

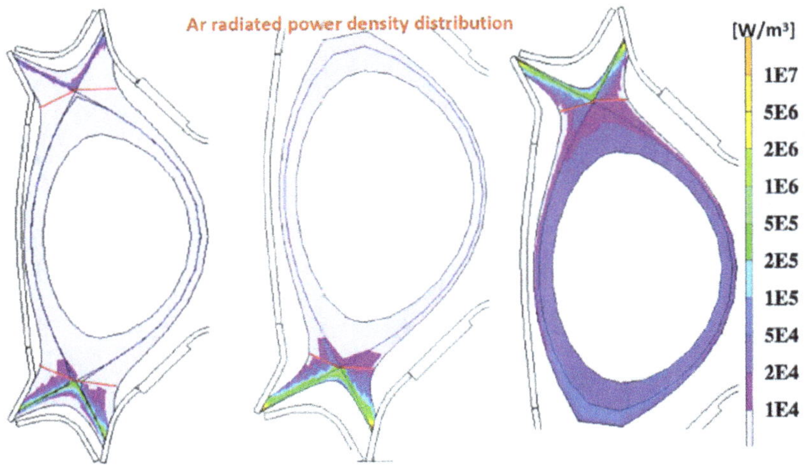

Figure 7.12. (a) Distribution of the radiated power density by argon (Ar) species in the computational region in modeling for double null, lower single null, and upper single null configurations. Reprinted from [26], with the permission of AIP Publishing.

penetration of impurities into the separatrix region. The effect of topology on impurity shielding can be understood by the secondary effect of a parallel SOL flow driven by perpendicular transport in tokamak devices.

7.3.2 Transport model

Impurity transport across the LCFS is an interesting issue because any impurities traveling outward undergo a sudden change in the background magnetic field condition from a closed-flux surface to an open magnetic field. The impurity transport across the LCFS is different from both the core impurity transport, where the gradient is determined by the balance between the diffusion and convection, and the SOL impurity transport, where the gradient is determined by the balance between friction force, thermal force, and electrostatic force. Once an impurity inside the LCFS reaches the boundary by perpendicular transport, it immediately escapes to the divertor by parallel transport. In contrast, an impurity outside the LCFS reaches the boundary and is immediately confined. The impurity density gradient is determined by the balance between the confinement inside the LCFS and confinement in the SOL region. Although this is an interesting and important issue, impurity transport across the LCFS has not been extensively modeled. In the impurity transport model inside the LCFS, inward flux across the LCFS is simply given as an impurity source to match the impurity concentration. In the impurity transport model in the SOL, inward flux across the LCFS is a sink of impurities at the LCFS. However, this inward flux across the LCFS is simply balanced to the outward flux across the LCFS (i.e., the sink is equal to the source at the LCFS) without solving the impurity transport across the LCFS. In a steady-state, both the impurity transport model inside the LCFS and in the SOL have zero net radial flux at the boundary of the LCFS. The impurity gradient to satisfy the boundary

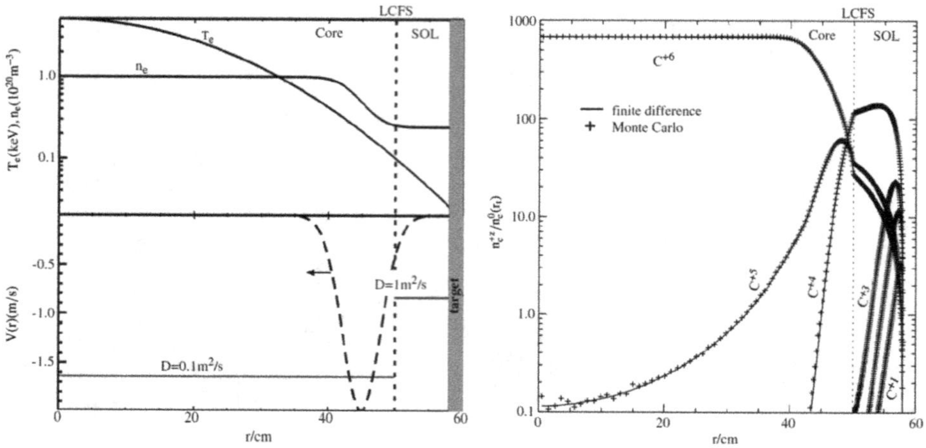

Figure 7.13. Radial profiles of electron density, n_e and T_e (top), and transport coefficients, V and D (bottom), as well as radial profiles of carbon density for each charge state normalized by carbon neutral density at the target (right). Reprinted from [28], Copyright (2013), with permission from Elsevier.

condition of zero net radial flux at the boundary of the LCFS is not determined in 3D edge Monte Carlo code such as the EMC3-Eirene code [27].

In order to determine the impurity density gradient at the LCFS, a numerical technique to couple the Monte Carlo code with a one-dimensional (1D) core model is necessary. Here, certain particular solutions to the 1D model are pre-calculated for given core plasma and transport coefficient profiles under specific boundary conditions at the SOL–core interface. Linear combination of these solutions yields a general solution, which is then translated into Monte Carlo language by formulating a so-called charge-state transition probability matrix. This matrix provides definitive boundary conditions at the SOL–core interface so that a self-consistent solution for both the SOL and core is attainable without the need for SOL–core iteration. Figure 7.13 shows radial profiles of electron density, n_e and T_e, transport coefficients, V and D, as well as radial profiles of carbon density for each charge state normalized by carbon neutral density at the target. The diffusion coefficient in the core is set to 0.1 m^2 s^{-1}, while the diffusion coefficient in the SOL is enhanced to 1 m^2 s^{-1}. The inward convection velocity is localized with a peak value of -2 m s^{-1} just inside the LCFS to reproduce the strong edge inward pinch and sharp gradient near the plasma periphery. The carbon impurity with a low charge state of C^{+1}, C^{+2}, C^{+3}, and C^{+4} is in the SOL region, and exhibits the so-called impurity shell, where the location is mainly determined by the plasma temperature. In contrast, the radial profile of the fully ionized carbon of C^{+6} is determined by the transport in the plasma, except for in the low-temperature region of the SOL. The gradient of fully ionized carbon has a very sharp density gradient near the plasma periphery, which is the result of the strong inward pinch in the transport model. The finite density gradient at the LCFS is determined by coupling the Monte Carlo code with the 1D core transport model.

7.3.3 Experimental results

In experiment, there have been many studies which have focused on core impurity transport and SOL impurity transport. However, as aforementioned, there have been few experiments regarding the impurity transport across the LCFS. In a recent study, the density gradients of hydrogen (H^+) and helium (He^{2+}) were measured with charge-exchange spectroscopy using a two-wavelength spectrometer in order to study the transport across the LCFS [19]. In the experiment, He impurity was selected to avoid the effects of atomic processes on impurity transport. Here, the SOL temperature was high enough to ionize the He to fully stripped ions. Figure 7.14 shows the radial profiles of H, He, and the helium-to-hydrogen ratio. The density gradient of He is much larger than the density gradient of H at the LCFS ($r_{\rm eff}/a_{99} = 1$). The radial profile of the helium-to-hydrogen ratio shows a sharp increase at the LCFS, which indicates the better particle confinement of He at the LCFS. In the core region, the H profile is peaked while the He profile is flat, which indicates that the H confinement is better than the He transport. This is in contrast to the He gradient being larger than the H gradient at the LCFS.

As seen in figure 7.13(d), the normalized density gradient (a/L_n) of He is 2 times larger than the normalized density gradient of H. As the He fraction (and electron density) are increased, the density gradient slightly decreases. Please note that the electron density also increases as the He fraction is increased because more He puff is added to increase the He fraction. This experiment demonstrates the importance of the study of impurity transport at the LCFS, because this impurity transport differs significantly from core impurity transport. The difference of transport between impurity and bulk ions can be understood as the difference in the ionization state of the source. The source of the bulk ions is neutral and can penetrate into the

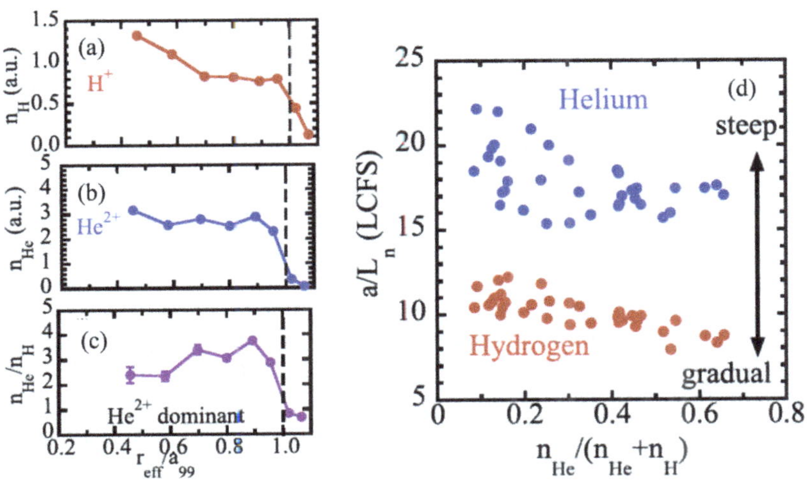

Figure 7.14. Radial profiles of (a) hydrogen (H^+), (b) helium (He^{2+}), and (c) helium-to-hydrogen ratio as well as (d) the gradients of H and He as a function of He fraction in the plasma, where L_n is the density scale length of H and He and a is the minor radius. Reproduced from [19].

plasma with thermal velocity regardless of the magnetic topology. In contrast, the impurity ions are partially ionized at the SOL and penetrate into the plasma through the LCFS. Therefore, the penetration is sensitive to changes in the magnetic topology across the LCFS, where the magnetic flux surface is nested with a closed magnetic field inside the LCFS, while the magnetic field is open and may be connected to a divertor or limiter configuration outside the LCFS. The larger gradient of He impurity than the bulk ions at the LCFS is unfavorable characteristic of He transport from the point of view of He ash exhaust. These observations are consistent with the sharp impurity density gradient predicted by the impurity transport model described in section 7.3.2.

References

[1] Bardóczi L, Rhodes T L, Carter T A, Crocker N A, Peebles W A and Grierson B A 2016 Non-perturbative measurement of cross-field thermal diffusivity reduction at the O-point of 2/1 neoclassical tearing mode islands in the DIII-D tokamak *Phys. Plasmas* **23** 052507

[2] Bardóczi L, Rhodes T L, Banon Navarro A, Sung C, Carter T A and La Haye R J *et al* 2017 Multi-field/-scale interactions of turbulence with neoclassical tearing mode magnetic islands in the DIII-D tokamak *Phys. Plasmas* **24** 056106

[3] Ida K, Kobayashi T, Ono M, Evans T E, McKee G R and Austin M E 2018 Hysteresis relation between turbulence and temperature modulation during the heat pulse propagation into a magnetic island in DIII-D *Phys. Rev. Lett.* **120** 245001

[4] Inagaki S, Tamura N, Ida K, Nagayama Y, Kawahata K and Sudo S *et al* 2004 Observation of reduced heat transport inside the magnetic island O point in the large helical device *Phys. Rev. Lett.* **92** 055002

[5] Ida K, Kobayashi T, Evans T E, Inagaki S, Austin M E and Shafer M W *et al* 2015 Self-regulated oscillation of transport and topology of magnetic islands in toroidal plasmas *Sci. Rep.* **5** 16165 EP

[6] Ida K, Kamiya K, Isayama A and Sakamoto Y 2012 Reduction of ion thermal diffusivity inside a magnetic island in JT-60U tokamak plasma *Phys. Rev. Lett.* **109** 065001

[7] Jahns G L, Ejima S, Groebner R J, Brooks N H, Fisher R K and Hsieh C L *et al* 1982 Dynamic behaviour of intrinsic impurities in Doublet III discharges *Nucl. Fusion* **22** 1049–59

[8] Ida K, Fonck R J, Hulse R A and Leblanc B 1986 Some effects of MHD activity on impurity transport in the PBX tokamak *Plasma Phys. Control. Fusion* **28** 879–95

[9] Weller A, Cheetham A D, Edwards A W, Gill R D, Gondhalekar A and Granetz R S *et al* 1987 Persistent density perturbations at rational-q surfaces following pellet injection in the Joint European Torus *Phys. Rev. Lett.* **59** 2303–6

[10] Tamura N, Liu Y, Iwama N, Sudo S, Khlopenkov K V and Kostrioukov A Y *et al* 2009 Confinement property of tracer impurity particle inside a static magnetic island O-point of Large Helical Device *J. Plasma Fusion Res. SERIES* **8** 975–80 (https://www.jspf.or.jp/JPFRS/PDF/Vol8/jpfrs2009_08-0975.pdf)

[11] Ida K, Inagaki S, Suzuki Y, Sakakibara S, Kobayashi T and Itoh K *et al* 2013 Topology bifurcation of a magnetic flux surface in magnetized plasmas *New J. Phys.* **15** 013061

[12] Finken K H, Abdullaev S S, Biel W, de Bock M F M, Busch C and Farshi E *et al* 2004 The dynamic ergodic divertor in the TEXTOR tokamak: plasma response to dynamic helical magnetic field perturbations *Plasma Phys. Control. Fusion* **46** B143–55

[13] Tokar M Z, Lasaar H, Mandl W, Hess W R and Michelis C D 1997 Modelling of plasma and impurity behaviour in a tokamak with a stochastic layer *Plasma Phys. Control. Fusion* **39** 569–689

[14] Telesca G, Delabie E, Schmitz O, Brezinsek S, Finken K H and von Hellermann M *et al* 2009 Carbon transport in the stochastic magnetic boundary of TEXTOR *J. Nucl. Mater.* **390–391** 227–31

[15] Breton C, Michelis C D, Mattioli M, Monier-Garbet P, Agostini E and Fall T *et al* 1991 Plasma decontamination during preliminary ergodic divertor experiments in TORE SUPRA *Nucl. Fusion* **31** 1774–81

[16] Chowdhuri M B, Morita S, Kobayashi M, Goto M, Zhou H and Masuzaki S *et al* 2009 Experimental study of impurity screening in the edge ergodic layer of the Large Helical Device using carbon emissions of C III to C VI *Phys. Plasmas* **16** 062502

[17] Ida K, Yoshinuma M, Tsuchiya H, Kobayashi T, Suzuki C and Yokoyama M *et al* 2015 Flow damping due to stochastization of the magnetic field *Nat. Comm.* **6** 5816 EP

[18] Morita S, Dong C F, Kobayashi M, Goto M, Huang X L and Murakami I *et al* 2013 Effective screening of iron impurities in the ergodic layer of the Large Helical Device with a metallic first wall *Nucl. Fusion* **53** 093017

[19] Ida K, Yoshinuma M, Goto M, Schmitz O, Dai S and Bader A *et al* 2016 Helium transport in the core and stochastic edge layer in LHD *Plasma Phys. Control. Fusion* **58** 074010

[20] Ida K, Yoshinuma M, Wieland B, Goto M, Nakamura Y and Kobayashi M *et al* 2015 Measurement of radial profiles of density ratio of helium to hydrogen ion using charge exchange spectroscopy with two-wavelength spectrometer *Rev. Sci. Instrum.* **86** 123514

[21] Kamiya K, Ida K, Yoshinuma M, Suzuki C, Suzuki Y and Yokoyama M 2012 Characterization of edge radial electric field structures in the Large Helical Device and their viability for determining the location of the plasma boundary *Nucl. Fusion* **53** 013003

[22] Conway G D, Angioni C, Ryter F, Sauter P and Vicente J 2011 Mean and oscillating plasma flows and turbulence interactions across the *L–H* confinement transition *Phys. Rev. Lett.* **106** 065001

[23] LaBombard B, Rice J E, Hubbard A E, Hughes J W, Greenwald M and Irby J *et al* 2004 Transport-driven Scrape-Off-Layer flows and the boundary conditions imposed at the magnetic separatrix in a tokamak plasma *Nucl. Fusion* **44** 1047–66

[24] Ida K and Rice J E 2014 Rotation and momentum transport in tokamaks and helical systems *Nucl. Fusion* **54** 045001

[25] Ryter Fthe H-Mode Database Working Group 1996 Results from the ITER H-mode threshold database *Plasma Phys. Control. Fusion* **38** 1279–82

[26] Xiang L, Guo H, Wischmeier M, Wu Z, Wang L and Duan Y *et al* 2017 Investigation of the effects of impurity seeding under different magnetic configurations in L-mode plasma in EAST tokamak *Phys. Plasmas* **24** 092514

[27] Feng Y, Sardei F, Kisslinger J, Grigull P, McCormick K and Reiter D 2004 3D edge modeling and island divertor physics *Contrib. Plasma Phys.* **44**

[28] Feng Y, Lunt T, Sardei F and Zha X 2013 Implicit coupling of impurity transport at the SOL-core interface *Comput. Phys. Comm.* **184** 1555–61

Chapter 8

Control of impurity accumulation

This chapter discusses methods of suppressing or mitigating the accumulation of impurities in magnetically confined plasmas. As mentioned in previous chapters, one of the critical milestones for acquiring stable fusion power in future fusion reactors based on the concept of magnetic confinement is to appropriately control the amount of impurities in the magnetically confined plasma. Controlling the impurities here does not simply mean keeping the impurity concentration in the magnetically confined plasma as low as possible. This is because impurities must be introduced into the plasma in a magnetic confinement fusion reactor in order to protect the plasma-facing components (PFCs) from tremendous heat loads. It would thus be counter-productive to control the impurities by eradicating them completely. In other words, we need to control the impurities in a magnetic confinement fusion reactor in such a way that they are appropriately distributed and in appropriate amounts. From this point of view, the issue of impurity accumulation, in which impurities accumulate progressively at the plasma center, represents an inadequate control with regards to impurity distribution and quantity.

A particular problem in impurity accumulation is accumulating elements with large atomic numbers (Z), as described in chapter 1.1. To avoid the accumulation of high-Z impurities in magnetically confined plasmas, the first thing to do is to suppress the generation of high-Z impurities by reducing the interaction between the plasma and PFCs made of these elements, which are the origin of the impurities. The next thing to be done is to prevent, as much as is possible, the generated high-Z impurities from entering the core plasma. The above measures to avoid the accumulation of high-Z impurities have already been discussed in related chapters. In this chapter, we discuss how to suppress or mitigate impurity accumulation in the plasma, assuming that high-Z impurities are already present to some extent. One possible method to suppress or mitigate the accumulation of impurities in magnetically confined plasma toward the plasma center is to suppress or apply a counteracting force to the driving force of impurity transport toward the plasma center.

The convection terms in neoclassical impurity transport are significant in transporting the impurities toward the plasma center. Impurity accumulation can be suppressed by reducing the contribution of these convection terms, for example by reducing the asymmetry between the low- and high-field sides (or LFS and HFS) by reducing the toroidal rotation in the case of tokamak plasmas or by decreasing the electron density and changing the sign of the radial electric field from negative to positive in the case of helical plasmas. In addition, since one of the driving forces for impurity transport toward the plasma center is the inward pinch due to the density gradient of the peaked electron density profile, impurity accumulation can be suppressed by flattening the electron density profile. In tokamaks, the enhancement of temperature screening is also an effective measure to combat impurity accumulation. Increasing turbulent transport and enhancing the outward impurity flux so that it exceeds the contribution of the convection terms in neoclassical transport can also suppress impurity accumulation.

To this end, radio-frequency (RF) heating is a promising method to help facilitate the above techniques in actual plasma operation. In the following two sections, we discuss the effects of additional RF heating on impurity accumulation.

8.1 Impact of electron cyclotron resonance heating

One of the possible wave heating methods in fusion plasmas is electron cyclotron resonance heating (ECRH), which can heat electrons efficiently. When plasma is heated up with ECRH, the electron temperature increases; thus, the electron temperature gradients outside the locations of wave absorption also increase. An increment of the electron temperature gradient can enhance the associated turbulence mode, such as the trapped electron mode (TEM). This enhanced turbulent transport can mitigate the accumulation of impurities in the core region of the plasma. Figure 8.1 shows an example discharge of the modification of argon (Ar) impurity behavior with ECRH in the KSTAR tokamak [1]. Here, an Ar gas puff was performed at 2.0 s (with 20 ms duration) from the midplane on the outboard side. As shown by the black lines, it can be seen that the Ar gas puff did not primarily affect the main plasma parameters, such as electron density and temperature. The radiation power from the highly ionized Ar impurity was measured with a soft X-ray (SXR) array diagnostic system. After the Ar gas puff, the signal intensity of the normalized central SXR chord, shown in the bottom panel of figure 8.1, gradually increased after a while, reached its peak, and then gradually decreased. Finally, it decreased to the level before the gas puff. In comparison, when 350 kW ECRH was applied 0.1 s before the Ar gas puff, the signal intensity of the normalized central SXR chord reached its peak faster and had a slighter peak value than when ECRH was not applied. Figure 8.2 shows the transport analysis results for the discharges shown in figure 8.1. The transport analysis shown here was performed using the SANCO (Stand Alone Non-Corona) impurity transport code [2], and the atomic data necessary for reproducing the measured signal intensity were taken from the Atomic Data and Analysis Structure (ADAS) database [3]. The neoclassical transport coefficients are calculated with the NCLASS code [4], which is

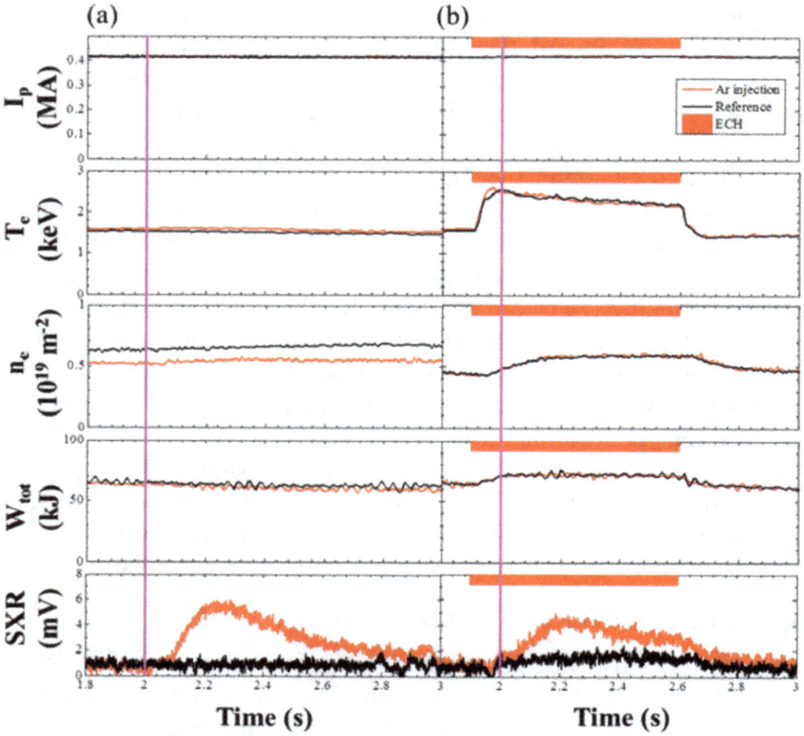

Figure 8.1. Comparison of the main parameters of KSTAR L-mode plasmas (a) without ECRH and (b) with ECRH. The vertical magenta lines represent the timing of the argon (Ar) gas puff. The data obtained with the Ar gas puff are shown with red lines, and as a reference, data obtained without Ar gas puff are shown with black lines. The red bars in the right frames denote the time during which ECRH was applied. Reproduced from [1].

a neoclassical code based on a fluid moment approach with a simplified collision operator. The profiles of electron density and temperature and ion temperature for the discharges without and with ECRH, which were used for the analysis, are shown in figures 8.2(a)–(c). As can be seen in figures 8.2(d) and (f), regarding diffusion coefficients, the neoclassical diffusion coefficients calculated with NCLASS (shown multiplied by 10) are found to be much smaller than the diffusion coefficients estimated with SANCO. With regard to convection velocities, the neoclassical convection velocities (also shown multiplied by 10) are near zero. Therefore, changes in neoclassical transport cannot explain the change in impurity behavior due to the application of ECRH in KSTAR low-confinement mode (L-mode) plasmas. Comparing the cases without ECRH and with ECRH, the ECRH significantly increased the diffusion coefficient in the region of $r/a < 0.25$, and the convective velocity around $r/a \sim 0.2$ became positive (i.e., outward). The results of linear gyrokinetic simulations using the GENE (Gyrokinetic Electromagnetic Numerical

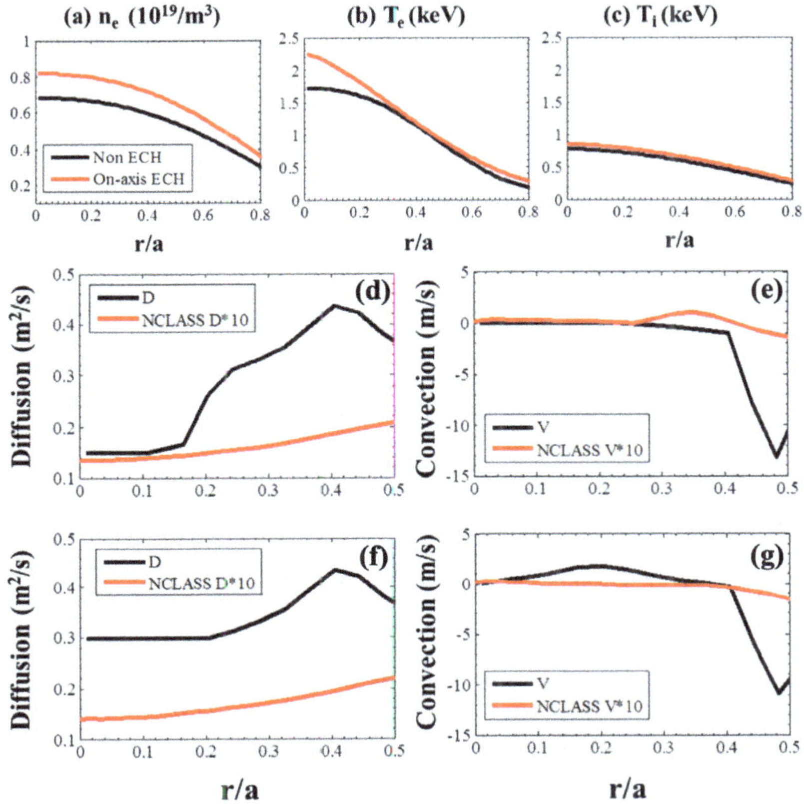

Figure 8.2. Comparison of the neoclassical transport coefficients (the diffusion coefficients (d) and (f) and convection velocity (e) and (g)), estimated with NCLASS and those calculated with SANCO for (d) and (e) the case without ECRH and (f) and (g) that with ECRH. In (d), (e), (f), and (g), the black lines represent the SANCO results and the red lines represent the NCLASS results. The NCLASS values are multiplied by 10. The normalized radial profiles of (a) electron density, (b) electron temperature, and (c) ion temperature used for the analysis are also shown. In (a)–(c), the black and red lines denote the cases without ECRH and those with ECRH, respectively. Reproduced from [1].

Experiment) code[1] suggest that the ECRH enhances TEM turbulence, which causes the outward convection velocity.

Another example, of the modification of molybdenum (Mo) impurity behavior with ECRH in L-mode plasmas of the EAST tokamak [5], is shown in figure 8.3. Here, the Mo, which originates from the first wall, is assumed to be the dominant intrinsic metallic impurity. The spatial profiles of Mo XXIX, Mo XXX, Mo XXXI, and Mo XXXII were measured in the device with space-resolved SXR and extreme ultraviolet (XEUV) spectrometers. When 450 kW ECRH was applied, the central electron density was changed from ~4.5 to ~3.8 × 10^{19} m^{-3}, and the central electron

[1] GENE Code homepage, available online: http://genecode.org.

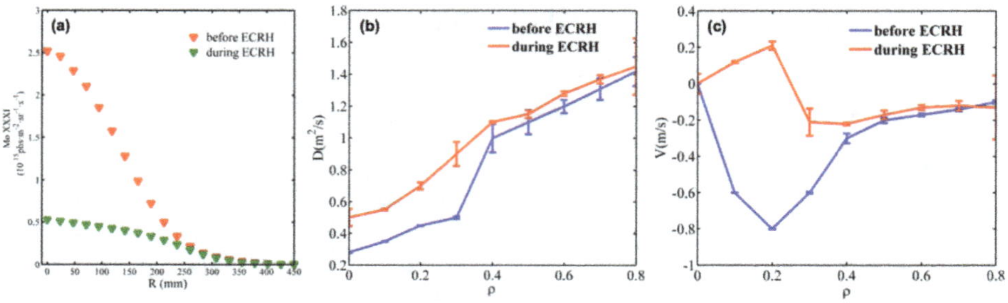

Figure 8.3. Radial profiles of (a) the sightline-integrated Mo XXXI intensity, (b) the diffusion coefficients, and (c) the convection velocities in ohmic EAST plasmas before and after adding ECRH. In (a), the red and green inverted triangles denote the data obtained before and during ECRH, respectively. In (b) and (c), the blue and red lines represent the data calculated before and during ECRH, respectively. Reprinted from [5], with the permission of AIP Publishing.

temperature from ~0.8 to ~1.7 keV. As shown in figure 8.3, the Mo XXXI intensity at $R < 250$ mm was drastically decreased by applying ECRH. The transport coefficients calculated using STRAHL [6] and the ADAS database, which were best fit for the reconstruction of the ion density profiles of Mo^{28+}, Mo^{29+}, Mo^{30+}, and Mo^{31+}, indicate that applying ECRH significantly changed the convection velocity from negative to positive in the core region where $r/a \leqslant 0.3$. Since the transport coefficients evaluated with the standard neoclassical theory were far from those calculated with STRAHL, it can be deduced that neoclassical transport does not play an important role in the experimental results with additional ECRH in the EAST L-mode plasmas. These facts are similar to the above-described experimental results obtained in the KSTAR L-mode plasmas. Therefore, they also support the idea that TEM turbulence is the main contributor to the change in convection velocity.

Next, we show results obtained in high-confinement mode (H-mode) tokamak plasmas. Figure 8.4 shows an example discharge of the modification of silicon (Si) impurity behavior with ECRH in H-mode plasmas of the ASDEX Upgrade tokamak [7]. Here, Si was injected into the ASDEX Upgrade plasmas by means of a laser blow-off (LBO) at 3.0 s. In this example, the Si LBO slightly perturbed the electron density and temperature at the edge region. As shown in the bottom frame of figure 8.4(a), the radiation power from the injected Si ions was measured with SXR cameras. In this experiment, ECRH with a power of 0.8 MW was additionally injected into the plasmas heated with 5 MW radial neutral beam injections (NBIs). The reasonable transport coefficients were calculated with STRAHL and ADAS using the profiles of measured electron density and temperature and the measured SXR intensity subtracted with background SXR intensity. For comparison, the effective heat diffusivity, $\chi_{\rm eff}$, and the neoclassical diffusivities of heat and Si impurity were calculated with the ASTRA [8] and NEOART codes [9], respectively. As a general trend, the diffusion coefficients decreased toward the center. At $r < 0.15$ m, where the power of the centrally focused ECRH (at $r = 0.075$ and 0.13 m) was absorbed, the diffusion coefficient in the plasma heated only by NBI is

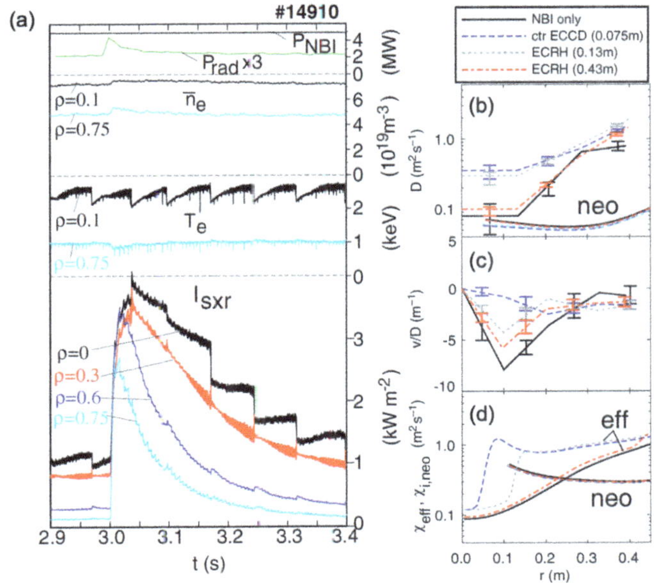

Figure 8.4. (a) Time history of the main parameters of ASDEX Upgrade tokamak H-mode plasmas with Si LBO at 3.0 s. Radial profiles of (b) the difussion coefficients, (c) drift parameter, and (d) effective heat diffusivity and (neoclassical) ion heat diffusivity for plasmas with various heating schemes. Reproduced from [7].

close to neoclassical. The diffusion coefficient in the plasmas additionally heated with centrally focused ECRH was significantly increased, while that at $r \sim 0.35$ m almost was not changed. This result is similar to the result obtained in the KSTAR L-mode plasmas. With regard to the drift parameter v/D (convection velocity/diffusion coefficient), all the cases with additional ECRH have smaller drift parameters compared to the NBI-only case. The case with additional counter-Electron Cyclotron Current Drive (ECCD) has the smallest drift parameter but does not reach a positive value.

Now, let us look at the case of a H-mode plasma with the same device. Figure 8.5 shows an example of the modification of Ar impurity behavior with ECRH in H-mode plasmas of the KSTAR tokamak [10]. The experimental scheme used for the KSTAR H-mode plasmas was the same as that for the L-mode plasmas, except for the applied ECRH power. When ECRH was applied, the electron density changed little, but the electron temperature increased with applied power, especially in the region $r/a < 0.6$. The ion temperature decreased in the central region for the 600 kW ECRH case, while it increased over the whole region for the 800 kW ECRH case. In contrast to the case of L-mode plasmas, the diffusion coefficients obtained using ADAS-SANCO did not change much with the application of ECRH. On the other hand, the convection velocity changed from a large negative to a small negative value depending on the applied ECRH power in the region of $r/a < 0.4$. Therefore, in H-mode plasmas of the KSTAR tokamak, the change (even still inside the negative regime) in convection velocity is considered to play a significant role in

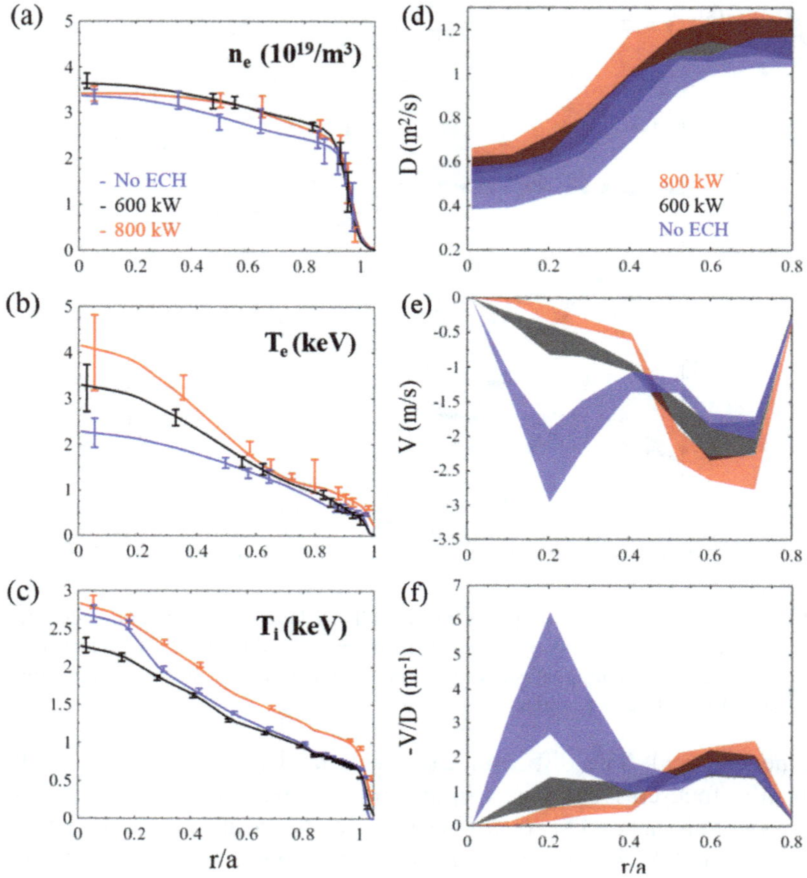

Figure 8.5. Normalized radial profiles of (a) electron density, (b) electron temperature, (c) ion temperature, (d) diffusion coefficient, (e) convection velocity, and (f) drift parameter for the three heating cases (without ECRH, blue; with 600 kW ECRH, black; with 800 kW ECRH, red). Reproduced from [10].

the change in impurity transport due to the application of ECRH. Here, the neoclassical transport coefficients, calculated with a local drift kinetic code, NEO [11], did not contribute to the change in the transport coefficients estimated with ADAS-SANCO. The role of turbulent transport in the results with ECRH was evaluated with a quasi-linear flux-tube gyrokinetic code, GKW [12]. The calculation results using GKW indicate that ion temperature gradient (ITG) modes were dominant in the case with ECRH, despite the heating effect. This may be because the conditions for TEM dominance, $T_e \gg T_i$, and $R/L_{Te} > 2R/L_{Ti}$ or $L_{Ti}/L_{Te} > 2$, were unmet. In other words, to enhance the turbulence-driven transport, in particular outward turbulent impurity flux, efficiently in suppressing or mitigating impurity accumulation, the ratio of T_e to T_i, i.e., Q_e to Q_i, is an important factor in tokamaks.

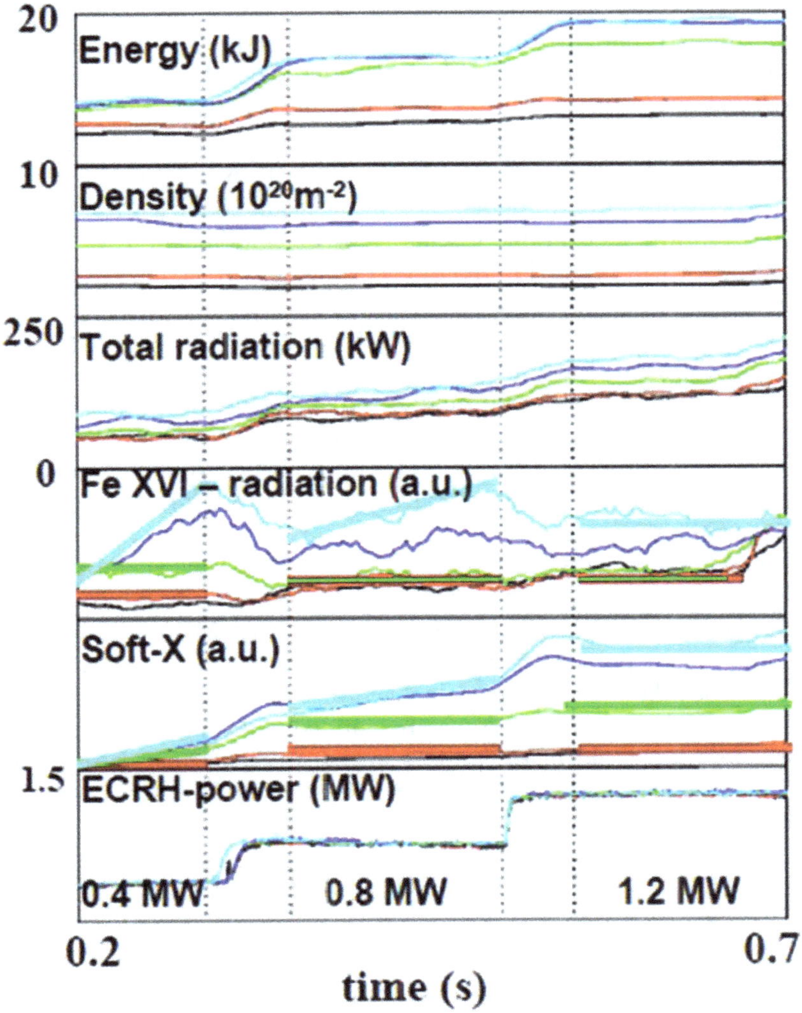

Figure 8.6. Time evolution of plasma parameters in W7-AS stellarator plasmas with different line-integrated electron density, but having the same stepwise increasing ECRH heating power. Reproduced from [13].

We now discuss examples from experiments adding ECRH in stellarators. Figure 8.6 shows an example discharge of the modification of iron (Fe) impurity behavior with ECRH in the W7-AS stellarator [13]. The time histories of plasmas with different line-integrated electron density are summarized in figure 8.6. In the W7-AS stellarator, Fe, which originates from the first wall, is a major intrinsic metallic impurity. In the figure, light blue lines denote the highest line-integrated electron density case, and blue lines denote the second-highest line-integrated electron density. During 0.4 MW ECRH, Fe XVI line radiation increased continuously in both the highest- and second-highest-density cases. Further, when the ECRH power was changed to 0.8 MW, Fe XV line radiation in the second-highest-density case remained at a certain level, but Fe XV line radiation in the highest-density case still increased constantly. Finally, with 1.2 MW ECRH, Fe XV line radiation, even in the highest-density case, remained at a constant level. In the W7-AS stellarator, this mitigation of impurity accumulation due to the addition of ECRH is considered to be due to an increase in diffusion coefficient, and turbulent transport could contribute to the increase in diffusion coefficient.

Figure 8.7 shows an example discharge of the modification of vanadium (V) impurity behavior with ECRH in the LHD heliotron [14]. Here, V was injected into LHD plasmas by means of a tracer-encapsulated solid pellet (TESPEL) at 3.95 s (50 ms before applying additional ECRH). When ECRH was not applied, the time evolution of the central electron temperature and line emissions from highly ionized V clearly indicate impurity accumulation, which causes the decrease in central electron temperature due to concentrated radiation. When 0.7 MW ECRH was applied after V injection, both V lithium (Li)-like and beryllium (Be)-like emissions decreased (but did not disappear completely), and the central electron temperature did not drop. Notably, after the ECRH was turned off, the V Be-like emissions increased significantly. This suggests that the 0.7 MW ECRH did not completely pump out the V impurity and that the residual V impurity began to accumulate in the plasma center again. When the ECRH power was 1.5 MW, the V Li-like and Be-like emissions were significantly reduced to almost zero compared to the 0.7 MW ECRH case. They did not increase even when the ECRH was turned off, suggesting that the V impurity was almost completely pumped out from the plasma. The profile analysis shows no significant differences in the gradients of electron density and temperature and ion temperature with the different ECRH power cases. As already explained, in stellarator devices, the radial electric field has a significant impact on the impurity transport. The radial electric fields measured with charge-exchange spectroscopy and those calculated by the DKES/PENTA code [15] did not change in the outer region ($r/a > 0.5$). Therefore, further detailed experiments (in particular, radial electric field measurements in the region of interest, where the ECRH power was absorbed locally) and analyses are needed to understand the mechanism of mitigating impurity accumulation in stellarator plasmas.

8.2 Impact of ion cyclotron resonance heating

Another possible wave heating method for fusion plasmas is ion cyclotron resonance heating (ICRH). ICRH has several advantages over ECRH with respect to the suppression or mitigation of impurity accumulation. The first is that either electrons, ions, or both can be efficiently heated, depending on the wave heating scheme and plasma composition. With respect to electron heating, as already discussed in the previous section, TEM turbulence may be driven by the enhancement of the electron temperature gradient. Accordingly, an increased ion temperature gradient for ion heating may drive ITG turbulence. Furthermore, optimizing the turbulence-driven particle flux may be possible by adjusting the electron-to-ion heating ratio. Second, the poloidal asymmetry of magnetically trapped ions heated by ICRH, as described

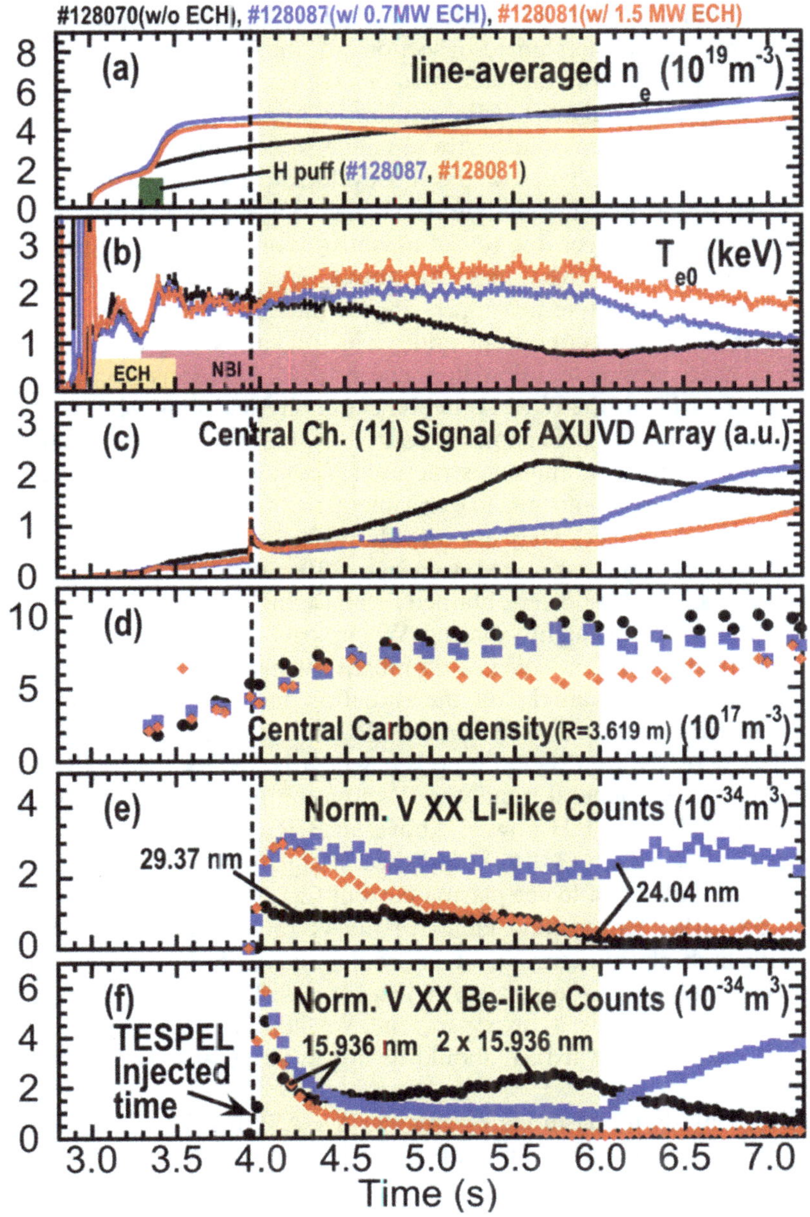

Figure 8.7. Time histories of the plasma parameters, (a) the line-averaged electron density, (b) the central electron temperature, (c) the chord-integrated signal intensity of channel 11 (from the center) of the absolute extreme ultraviolet photodiode array, (d) the central carbon density, and the normalized integrated counts for the V (e) Li-like and (f) Be-like emissions, for LHD plasmas with different heating schemes: without additional ECRH (black), with 0.7 MW (blue), and 1.5 WM (red) ECRH. Additional ECRH was applied from $t = 4.0$ s to $t = 6.0$ s, as indicated by the light brown hatched regions. Reprinted from [14], with the permission of AIP Publishing.

in chapter 5.3, can mitigate the poloidal asymmetry of heavy impurity ions due to centrifugal force. As also explained in section 5.3, the poloidal asymmetry of heavy impurity ions due to centrifugal force weakens the temperature screening effect and induces impurity accumulation. Minority ion heating by ICRH can weaken the poloidal asymmetry of heavy impurity ions due to centrifugal force, although the location of the resonance layer of the ion cyclotron wave must be appropriately selected. Third, the effective temperature, $T_{\text{eff}} = (T_{\|} + 2T_{\perp})/3$, may enhance the temperature screening effect due to the anisotropy of the temperature of minority ions heated by ICRH. Given that the above advantages are mixed, ICRH is expected to be very effective in suppressing or mitigating impurity accumulation, although there are certain conditions under which it can be applied. However, there is one point that requires caution in the use of ICRH. Ions accelerated by the potential generated near the ICRH antenna cause physical sputtering at the antenna [16], and the antenna itself becomes a source of impurity particles, which can be minimized by the choice of anntena structure [17]. Nevertheless, the effectiveness of ICRH heating has been confirmed in experiments with several magnetically confined plasma experiment devices. They are introduced below.

Figure 8.8 shows the estimated tungsten (W) radiation peaking factor as a function of applied ICRH power obtained in a baseline scenario in the JET tokamak [18]. Radiation from the highly ionized W impurity was measured with an SXR array diagnostic system. Since W is a dominant intrinsic impurity derived from the first wall of the JET tokamak, all the radiation measured with the SXR array diagnostic system is assumed to be from the W. In this data set, the RF power does not exceed 20% of the total heating power, which is mainly from NBI. As can be seen in figure 8.8, as a general trend, the W radiation peaking factor decreased with the increase in applied ICRH power. Above an ICRH power of 2.5 MW, plasmas with higher plasma current (3.0–4.0 MA) show a higher W radiation peaking factor, compared to plasmas with lower plasma current (2.5 MA). When the ICRH power was set to 5.0 MW, the peaking factor of the W radiation remained low even in plasmas with high plasma currents. The profile analysis indicates that the normalized electron temperature gradient R/L_{Te} increased with ICRH power. The resulting $R/L_{\text{n}} - 0.5R/L_{\text{T}}$, which is a proxy for the W flux, was changed from positive (inward) to negative (outward). It should be noted here that $n_{\text{i}} = n_{\text{e}}$ and $T_{\text{i}} = T_{\text{e}}$ are assumed.

A similar analysis for H-mode plasmas in the ASDEX Upgrade tokamak, but with a more detailed comparison of the effects of ECRH and ICRH on W accumulation, is shown in figure 8.9(a). Tungsten is a dominant intrinsic impurity derived from the first wall of the ASDEX Upgrade tokamak, similar to the JET tokamak. In this figure, the W density at the central ($\rho_\phi \sim 0.15$) and middle ($\rho_\phi \sim 0.5$) regions was evaluated from measurements with a grazing incidence spectrometer (GIW). As a comparison of the impact of ECRH and ICRH on the W accumulation in the H-mode plasmas of ASDEX Upgrade, experimental results with localized (near on-axis) ECRH, near on-axis ICRH, and broad ECRH (imitating an ICRH power absorption profile) are summarized in figure 8.9(a).

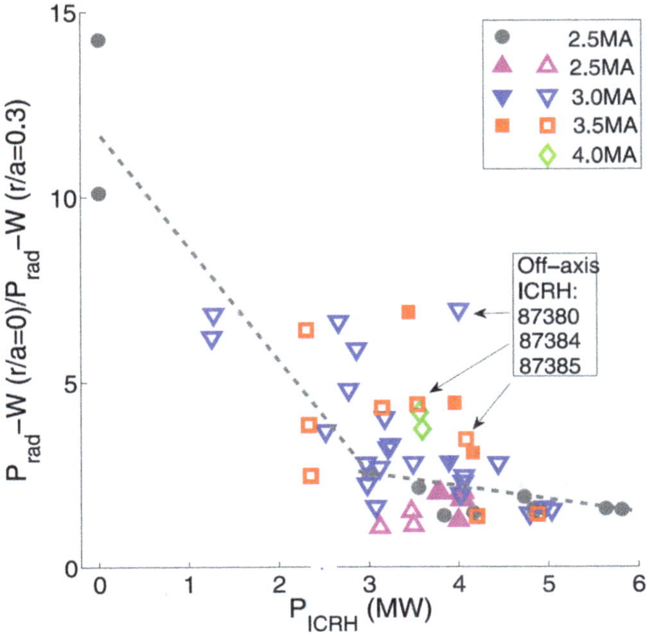

Figure 8.8. Tungsten (W) radiation peaking factor as a function of ICRH power with different plasma currents. Here, the W peaking factor is defined as $(P\text{rad} - W(r/a = 0))/(P\text{rad} - W(r/a = 0.3))$. Closed symbols denote discharges with low to medium gas rate ($1\text{–}2 \times 10^{22}$ el s^{-1}), and open symbols denote discharges with high gas rate ($2.3\text{–}5 \times 10^{22}$ el s^{-1}). Closed gray circle symbols represent low pumping discharges. The off-axis ICRH cases ($R_{\text{IC}} - R_{\text{mag}} \sim 0.40$ m) are emphasized. Reproduced from [18].

In the experiments with ECRH, to match the W source level with the experiments with ICRH, low-power ICRH was also used. As can be seen in figure 8.9(a), the estimated W density peaking factor decreased with the increase in RF heating power. The heating power required to achieve a flat W density profile is the smallest in the case of localized ECRH. For ICRH, it appears that more than twice as much heating power is required to achieve a similar flat W density profile. However, when the electron heating portion of ICRH heating is taken into account, the electron heating power needed to flatten the W density is approximately the same for all RF heating schemes. In other words, this series of experiments indicates that electron heating, rather than ion heating, plays an important role in suppressing the W accumulation. The relation between the W density peaking factor and the electron heat flux fraction inside $r/a < 0.25$ for each of the heating conditions is shown in figure 8.9(b). The electron heat flux fraction absorbed in the region of $r/a < 0.25$ was evaluated by power-balance analysis with TRANSP [19]. As can be seen from figure 8.9(b), there is a clear dependence of the W density peaking factor on the electron heat flux fraction in the core region. However, profile analysis for these discharges indicates that larger normalized ion temperature gradients R/L_{Ti} were obtained by efficient ion heating with ICRH as compared to the ECRH cases.

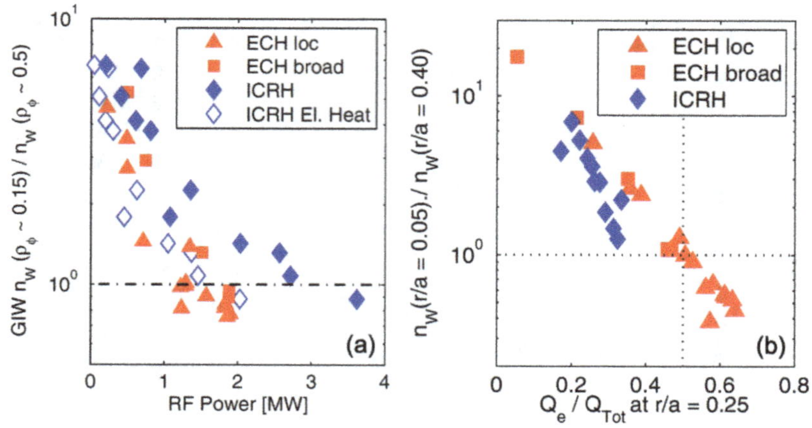

Figure 8.9. (a) Ratio of tungsten (W) density measured with a grazing incidence spectrometer (GIW) in between the central and middle regions as a function of total power of additional centrally localized ECRH (red triangle), centrally broadened ECRH (red square), and near on-axis ICRH (blue solid diamond). The open blue diamonds represent the electron heating fraction of the applied ICRH power. (b) W density peaking factor as a function of the electron heat flux fraction inside $r/a < 0.25$ for each of the heating conditions. Reproduced from [20].

Further, the normalized electron density gradient $R/L_{\rm n}$ decreased with RF heating power. Therefore, RF heating modifies the neoclassical convection, $R/L_{\rm n} - 0.5R/L_{\rm Ti}$), in a direction where impurity accumulation is suppressed or mitigated. Furthermore, with ICRH, the modification of such neoclassical convection is found to be stronger. From the results shown above, it can be seen that ICRH is effective in suppressing or mitigating the accumulation of impurities because it can modify both neoclassical and turbulent transport by heating not only electrons but also ions.

8.3 Other effects on impurity accumulation

In the previous section, we have shown how RF heating, such as ECRH and ICRH, contributes to suppressing or mitigating the accumulation of impurities in magnetically confined high-temperature plasmas. It is now known that neoclassical and turbulent transport each contribute significantly to the mechanisms of suppression or mitigation of impurity accumulation. The turbulence-driven impurity flux can be analytically expressed as [21]

$$\frac{R\Gamma_Z}{n_Z} = D_{nZ}\frac{R}{L_{nZ}} + D_{TZ}\frac{R}{L_{TZ}} + D_{uZ}u_Z' + V_{pZ}, \tag{8.1}$$

where D_{nZ}, D_{TZ}, D_{uZ}, and V_{pZ} represent the coefficients of impurity turbulent diffusion, thermo-diffusion, roto-diffusion, and pure convection, respectively. In addition, practically, these normalized coefficients $C_{TZ} = D_{TZ}/D_{nZ}$, $C_{uZ} = D_{uZ}/D_{nZ}$, and $C_{pZ} = RV_{pZ}/D_{nZ}$ for thermo-diffusion, roto-diffusion, and pure convection, respectively, can be used for comparison with quasi-linear calculations. For local descriptions of transport and trace impurities, the equation is linear, and, thus, the experimental and numerical investigation of each term's contribution is being studied. Here, we show an example of an experimental investigation of the possible contribution of turbulent impurity flux in the suppression of impurity accumulation, in particular one such contribution: roto-diffusion. Figure 8.10 shows an example discharge of the suppression of impurity accumulation with strong NBI heating observed in the LHD heliotron. In the LHD, impurity accumulation can be seen to occur with lower NBI power (9.5 MW), as indicated by the central plasma radiation and the Fe xxiii line radiation. As a result, the central electron temperature decreased significantly. The main intrinsic impurity in the LHD is carbon, originating from the LHD divertor plates. As can be seen in figures 8.10(e) and (g), when the impurity accumulation occurred in the LHD, carbon accumulation also appeared. Consequently, a peaked carbon density profile was formed. On the other hand, with higher NBI power (13 MW), the central plasma radiation and the Fe XXIII line radiation were kept at a very low level. The carbon density profile retained its shape, and the discharge was terminated as scheduled. Based on calculation with the DKES/

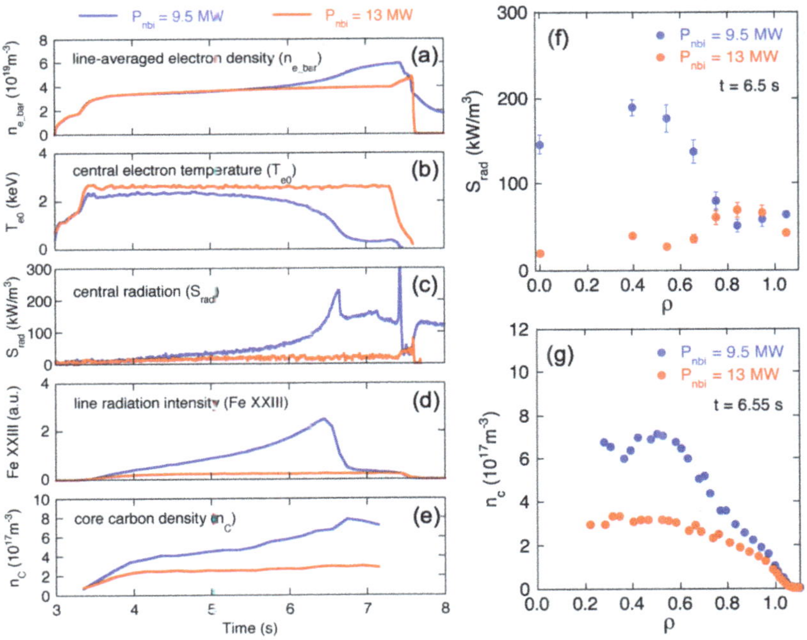

Figure 8.10. Time history of (a) line-averaged electron density, (b) central electron temperature, (c) central plasma radiation power, (d) Fe XXIII intensity, and (e) central carbon density of LHD plasmas heated with different NBI powers (9.5 MW, blue; 13 MW, red). Also shown are radial profiles of (f) plasma radiation power and (g) carbon density of LHD plasmas heated with different NBI powers. Reproduced from [23].

PENTA code, even in the plasma with higher NBI power, the sign of the radial electric field was still negative. Therefore, the standard neoclassical theory without an external torque input cannot explain the suppression of the impurity accumulation in the LHD plasma with higher NBI power.

In addition, in stellarator devices the strong tangential NBI heating can rotate the plasma. Thus, the contribution of toroidal plasma rotation on the suppression of impurity accumulation was investigated. Figure 8.11(a) shows the normalized carbon density gradient as a function of normalized toroidal rotation gradient obtained in LHD plasmas without impurity accumulation due to strong NBI heating. As can be seen in figure 8.11(a), the shape of the carbon density profile changed from peaked to hollow with the increase in the toroidal rotation gradient, i.e., there is a clear correlation between the carbon density gradient and the toroidal rotation gradient. A similar trend is also observed in the ASDEX Upgrade tokamak [22], which is shown in figure 8.11(b). In ASDEX Upgrade, the observation was performed for boron impurity. Gyrokinetic simulation with GKW exhibited a good (close to quantitative) agreement with the experimental results. However, gyrokinetic simulation studies of helium transport have failed to reproduce its characteristics. Due to their low collisionality, the contribution of turbulent transport is significant in fusion plasmas, even for heavy impurities (and even in stellarator plasmas). Therefore, it is necessary to accurately understand the characteristics of turbulent impurity transport, including for heavy impurities. As at present this research is still in its infancy, experimental and numerical studies on impurity transport should continue in the future.

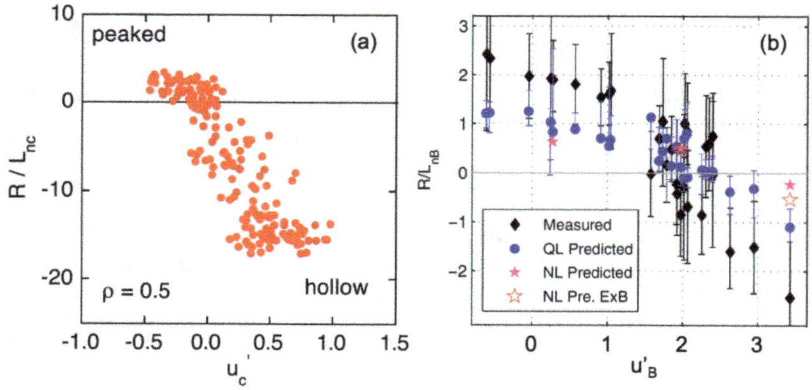

Figure 8.11. (a) Normalized carbon density gradient as a function of the normalized toroidal rotation gradient at $r/a = 0.5$ in the LHD heliotron, (b) comparison of the quasi-linear (QL) predicted and measured normalized boron density gradient dependence on the toroidal rotation gradient at $r/a = 0.5$ in the ASDEX Upgrade tokamak. The QL predictions are calculated from a spectrum with its peak at $k_\theta \rho_i = 0.3$, which is compared with non-linear (NL) simulations for selected points. Panel (a) reproduced from [23]. Panel (b) reproduced from [22].

References

[1] Hong J, Lee S H, Kim J, Seon C R, Lee S G and Park G Y *et al* 2015 Control of core argon impurity profile by ECH in KSTAR L-mode plasmas *Nucl. Fusion* **55** 063016

[2] Lauro-Taroni L 1994 *Impurity Transport of high performance discharges in JET* JET-P-4-29 INIS Report (https://inis.iaea.org/collection/NCLCollectionStore/_Public/28/032/28032669.pdf)

[3] Summers H P 2023 The ADAS User Manual, version 2.6, available from https://www.adas.ac.uk/

[4] Houlberg W A, Shaing K C, Hirshman S P and Zarnstorff M C 1997 Bootstrap current and neoclassical transport in tokamaks of arbitrary collisionality and aspect ratio *Phys. Plasmas* **4** 3230–42

[5] Shen Y, Lyu B, Zhang H, Li Y, Fu J and Vogel G *et al* 2019 Suppression of molybdenum impurity accumulation in the core using on-axis electron cyclotron resonance heating in EAST *Phys. Plasmas* **26** 032507

[6] Behringer K 1987 *Description of the impurity transport code STRAHL* JET-R(87)08 JET Joint Undertaking Report

[7] Dux R, Neu R, Peeters A G, Pereverzev G, Mück A and Ryter F *et al* 2003 Influence of the heating profile on impurity transport in ASDEX Upgrade *Plasma Phys. Control. Fusion* **45** 1815

[8] Pereverzev G V and Yushmanov P N 2002 *ASTRA—Automated System for TRansport Analysis* IPP 5/98 Max-Planck-Institut Für Plasmaphysik IPP-Report

[9] Peeters A G 2000 Reduced charge state equations that describe Pfirsch–Schlüter impurity transport in tokamak plasma *Phys. Plasmas* **7** 268–75

[10] Hong J, Henderson S S, Kim K, Seon C R, Song I and Lee H Y *et al* 2017 Modification of argon impurity transport by electron cyclotron heating in KSTAR H-mode plasmas *Nucl Fusion* **57** 036028

[11] Belli E A and Candy J 2008 Kinetic calculation of neoclassical transport including self-consistent electron and impurity dynamics *Plasma Phys. Control. Fusion* **50** 095010

[12] Peeters A G, Camenen Y, Casson F J, Hornsby W A, Snodin A P and Strintzi D *et al* 2009 The nonlinear gyro-kinetic flux tube code GKW *Comput. Phys. Commun.* **180** 2650–72

[13] Burhenn R, Feng Y, Ida K, Maassberg H, McCarthy K J and Kalinina D *et al* 2009 On impurity handling in high performance stellarator/heliotron plasmas *Nucl. Fusion* **49** 065005

[14] Tamura N, Suzuki C, Satake S, Nakamura Y, Nunami M and Funaba H *et al* 2017 Observation of the ECH effect on the impurity accumulation in the LHD *Phys. Plasmas* **24** 056118

[15] Spong D A 2005 Generation and damping of neoclassical plasma flows in stellarators *Phys. Plasmas* **12** 056114

[16] Perkins F W 1989 Radiofrequency sheaths and impurity generation by ICRF antennas *Nucl. Fusion* **29** 583

[17] Bobkov V, Braun F, Dux R, Herrmann A, Faugel H and Fünfgelder H *et al* 2016 First results with 3-strap ICRF antennas in ASDEX Upgrade *Nucl. Fusion* **56** 084001

[18] Goniche M, Dumont R J, Bobkov V, Buratti P, Brezinsek S and Challis C *et al* 2017 Ion cyclotron resonance heating for tungsten control in various JET H-mode scenarios *Plasma Phys. Control. Fusion* **59** 055001

[19] Breslau J, Gorelenkova M, Poli F, Sachdev J, Pankin A and Perumpilly G *et al* 2018 *TRANSP computer software* Available from: https://www.osti.gov/biblio/1489900

[20] Angioni C, Sertoli M, Bilato R, Bobkov V, Loarte A and Ochoukov R *et al* 2017 A comparison of the impact of central ECRH and central ICRH on the tungsten behaviour in ASDEX Upgrade H-mode plasmas *Nucl. Fusion* **57** 056015

[21] Angioni C 2021 Impurity transport in tokamak plasmas, theory, modelling and comparison with experiments *Plasma Phys. Control. Fusion* **63** 073001

[22] Casson F J, McDermott R M, Angioni C, Camenen Y, Dux R and Fable E *et al* 2013 Validation of gyrokinetic modelling of light impurity transport including rotation in ASDEX Upgrade *Nucl. Fusion* **53** 063026

[23] Nakamura Y, Tamura N, Yoshinuma M, Suzuki C, Yoshimura S and Kobayashi M *et al* 2017 Strong suppression of impurity accumulation in steady-state hydrogen discharges with high power NBI heating on LHD *Nucl. Fusion* **57** 056003

www.ingramcontent.com/pod-product-compliance
Lightning Source LLC
LaVergne TN
LVHW062240200926
848484LV00006B/1577